Managing E-Learning:
Design, Delivery, Implementation and Evaluation

Badrul H. Khan
George Washington University, USA

 **Information Science Publishing**

Hershey • London • Melbourne • Singapore

Acquisitions Editor:	Renée Davies
Development Editor:	Kristin Roth
Senior Managing Editor:	Amanda Appicello
Managing Editor:	Jennifer Neidig
Copy Editor:	Jennifer Young
Typesetter:	Jennifer Neidig and Sara Reed
Cover Design:	Lisa Tosheff
Printed at:	Yurchak Printing Inc.

Published in the United States of America by
      Information Science Publishing (an imprint of Idea Group Inc.)
      701 E. Chocolate Avenue, Suite 200
      Hershey PA 17033
      Tel: 717-533-8845
      Fax: 717-533-8661
      E-mail: cust@idea-group.com
      Web site: http://www.idea-group.com

and in the United Kingdom by
      Information Science Publishing (an imprint of Idea Group Inc.)
      3 Henrietta Street
      Covent Garden
      London WC2E 8LU
      Tel: 44 20 7240 0856
      Fax: 44 20 7379 3313
      Web site: http://www.eurospan.co.uk

Library of Congress Cataloging-in-Publication Data

Khan, Badrul Huda.
  Managing e-learning : design, delivery, implementation, and evaluation / Badul Khan.
    p. cm.
  Summary: "This book provides readers with a broad understanding of the emerging field of e-learning and also advises readers on the issues that are critical to the success of a meaningful e-learning environment"--Provided by publisher.
  Includes bibliographical references and index.
  ISBN 1-59140-634-X (hard cover) -- ISBN 1-59140-635-8 (soft cover) -- ISBN 1-59140-636-6 (ebook)
  1. Distance education--Computer-assisted instruction--Management. 2. Web-based instruction--Management. I. Title.
  LC5803.C65K53 2005
  371.3'58--dc22
                            2005004526

British Cataloguing in Publication Data
A Cataloguing in Publication record for this book is available from the British Library.

Dedication

This book is dedicated to my late parents:

Mr. Lokman Khan Sherwani &
Mrs. Shabnom Khanam Sherwani
of
Khan Manzil, Pathantooly, Chittagong, Bangladesh

Managing E-Learning:
Design, Delivery, Implementation and Evaluation

Table of Contents

Preface

Growing up in Bangladesh during the 1970s, I used to dream about having access to the well-designed learning resources that were only available to students in industrial countries. In the '70s it was unthinkable that we might have equal access to those resources. In the '90s, it became a reality. We are blessed with the emergence of the Internet's World Wide Web as one of the most important economic and democratic mediums of learning and teaching at a distance. The Internet has become an increasingly powerful, global, interactive and dynamic medium for sharing information. The Internet provides an open, dynamic and flexible learning environment with implications for countless applications with respect to education and training. Internet technologies provide an opportunity to develop new learning experiences for students that have not been possible before.

With the advent of the Internet and online learning methodologies and technologies, providers of education and training are creating e-learning materials to fulfill the demand. Online learning is becoming more and more accepted in workplace. Institutions are investing heavily in the development and deployment of online programs. Academic institutions, corporations and government agencies worldwide are increasingly using the Internet and digital technologies to deliver instruction and training.

What does it take to create a successful e-learning environment for diverse learners? A successful e-learning system involves a systematic process of planning, design, development, evaluation and implementation to create an online environment where learning is actively fostered and supported. In order for an e-learning system to be successful, it must be meaningful to all stakeholder groups, including learners, instructors, support services staff and the institution.

An e-learning system is meaningful to *learners* when it is easily accessible, well-designed, learner-centered, affordable and efficient, flexible and has a facilitated learning environment. When learners display a high level of participation and success in meeting a course's goals and objectives, this can make e-learning meaningful to *instructors*. In turn, when learners enjoy all available support services provided in the course without any interruptions, it makes *support services staff* happy as they strive to provide easy-to-use, reliable services. Finally, an e-learning system is meaningful to *institutions* when it has a sound return-on-investment (ROI), a moderate to high level of learner satisfaction with both the quality of instruction and all support services and a low drop-out rate.

To create a meaningful e-learning environment for diverse learners, we must explore various important issues encompassing various dimensions of e-learning environment.

The purpose of this book is to provide you with a broad understanding of the emerging field of e-learning and also to advise you on the issues that are critical to the success of a meaningful e-learning environment. It walks you through the various factors critical to developing, evaluating and implementing e-learning. Throughout the book critical e-learning factors are presented as questions that you can ask yourself when planning, designing, evaluating and implementing e-learning modules, courses and programs.

In this book, I present various critical issues of open, flexible and distributed e-learning environments with "A Framework for E-Learning." The seeds for the e-Learning Framework began germinating with the question, "What does it take to provide flexible learning environments for learners worldwide?" After the publication of my book, Web-Based Instruction (Educational Technology Publications, 1997), readers began e-mailing me to ask whether I could point them to a Web-based course that I thought was truly Web-based instruction, as I defined it in the book. Unfortunately, I did not have an answer for them. In 1997, the Web was used more for presenting information than for designing instruction; people were still experimenting with the Web. However, the rapid growth of e-learning over the next few years provided a rich climate for further exploration of this question.

Since 1997, I have been communicating with learners, instructors, administrators and technical and other support services staff involved in e-learning, in both academic and corporate settings, all over the world. I have researched e-learning issues discussed in professional discussion forums, newspapers, magazines and journals, and I have designed and taught online courses. Also, as the editor of Web-Based Training (Educational Technology Publications, 2001) and the forthcoming Flexible Learning (Educational Technology Publications), I have had the opportunity to work closely on critical e-learning issues with more than 100 authors worldwide who contributed chapters to these books.

Through these activities, I found that numerous factors help to create a meaningful learning environment, and many of these factors are systemically interrelated and interdependent. A systemic understanding of these factors can help us create meaningful distributed learning environments. I clustered these factors into eight categories: institutional, management, technological, pedagogical, ethical, interface design, resource support, and evaluation. I found these eight categories to be logically comprehensive and empirically the most useful dimensions for e-learning environments. With these eight dimensions, I developed "A Framework for E-Learning." The framework is reviewed by researchers and practitioners from various countries (http://BooksToRead.com/framework/ #acknowledgement), and I am indebted to them for their insightful comments that truly improved the framework.

The book is organized according to the eight dimensions of the framework. These eight dimensions would blend together during the planning, design, development, implementation and evaluation process leading to a quality e-learning system. Chapter 1 introduces e-learning as an open, flexible and distributed learning environment; how its various learning features can be designed to address critical issues encompassing the various dimension of e-learning environment. Each dimension of e-learning environment is discussed in detail in a separate chapter. Institutional, management, technological, pedagogical, ethical, interface design, resource support and evaluation issues of e-learning are discussed respectively in Chapter 2 through Chapter 9. Numerous factors discussed in Chapter 2 through Chapter 9 should give you a comprehensive picture of e-learning and should also help you think through every aspect of what you are doing during various steps of e-learning process.

Within the scope of this book, through the lens of the e-Learning Framework, only several critical items or questions related to each dimension of the e-learning environment are presented as examples at the end of Chapter 2 through Chapter 9; they (i.e., items) are thus by no means complete. However, there are myriad of important items or questions encompassing the various dimensions of e-learning environment that need to be explored. Please note that each e-learning project is unique. I encourage you to identify as many issues as possible for your own e-learning project by using the framework. One way to identify critical issues is by putting each stakeholder group (such as learner, instructor, support staff, etc.) at the center of the framework and raising issues along the eight dimensions of the e-learning environment. This way you can identify many critical issues that can help create meaningful e-learning environment for that particular group. By repeating the same process for other stakeholder groups, you can generate a comprehensive list of issues for your e-learning project.

For each section of text in Chapter 2 through Chapter 9 there are corresponding checklist items. I highly recommend that you review checklist items while reading text in each section of the book. Checklist items elaborate issues covered in each section of the book.

You may be thinking — how many issues do I have to address? How many issues are necessary? It depends on the goals and scope of your project. The more e-learning issues you explore and address, the more meaningful and supportive a learning environment you help to create for your target population.

Designing open, flexible and distributed e-learning systems for diverse learners is challenging; however, as more and more institutions offer e-learning to learners worldwide, we will become more knowledgeable about what works and what does not work. We should try our best to accommodate the needs of stakeholder groups by asking as many critical questions as possible along the eight dimensions of e-learning environment. The number and types of questions may vary based on each unique e-learning system. Given our specific e-learning contexts, we may not be able to address all the critical issues within the eight dimensions of e-learning. We should find ways to address them with the best possible means that we can afford. It is important to ask many questions as possible during the planning period of e-learning design.

In Chapter 2 through Chapter 9, I discuss as many critical issues as possible with examples of relevant e-learning cases from around the world. The world of e-learning is constantly changing and evolving. To keep you up to date with resources, FAQs, strategies, best practice examples and any change of addresses for chapter-related Web sites and other corrections, I maintain a Web site at: http:// BooksToRead.com/elearning.

Who can benefit from this book? I believe a wide range of people can use the book:

Instructors can use this book as textbook or reference work in courses dealing with topics such as distance education, technology in education, designing online education, program evaluation of online education, designing blended-learning, e-learning, Web-based instruction, distributed learning, computers in education, multimedia, educational technology, instructional technology, educational telecommunications, teacher training, instructional design, corporate training and so on. This book can also be used as a supplemental to other educational and information technology related courses.

In addition, instructors, teachers, trainers, training managers, distance education specialists, e-learning specialists, virtual education specialists, e-learning project managers, instructional designers, corporate education specialists, human resources specialists, performance technologists, educational technology coordinators, media specialist, Webmasters, writers/editors and technical support staff can use this book to plan, design, evaluate and implement e-learning modules, courses and programs.

Virtual/corporate university designers can use this book to plan, design, evaluate and implement corporate/virtual universities.

School administrators, higher education administrators, department of education staff, ministry of education staff, virtual and corporate university administrators, human resources managers and consultants can use this book to develop strategic plans for designing, evaluating and implementing e-learning initiatives.

Anyone contemplating a career in training and development, curriculum planning and Internet applications can use this book to learn about e-learning design strategies.

Finally, I hope that various e-learning issues included in this book will help you understand all aspects of e-learning environment and provide valuable guidance in creating e-learning and blended learning experience for your target audience. I would appreciate hearing your comments regarding this book.

Instructions to Use This Book

- If you are designing or evaluating e-learning degree/certificate programs, distance education programs and virtual universities, it is recommended that you start with Chapter 2.

- If you are designing or evaluating e-learning courses and lessons, it is recommended that you start with Chapter 3. Please note that some issues in Chapter 2 may not be relevant to your projects.

- For each section of text in Chapter 2 through Chapter 9 there are corresponding checklist items. It is *highly recommended* that you review checklist items while reading text in each section of the book. Checklist items elaborate issues covered in each section of the book.

Badrul H. Khan
bhk@BooksToRead.com
http://BadrulKhan.com/khan

For a list of glossary terms on e-learning, visit http//:BooksToRead.com/ elearning/glossary.htm and for a list of e-learning resources, visit http:// BooksToRead.com/elearning/resources.

Acknowledgments

This book owes much to the encouragement and assistance of many people. I would like to thank my graduate students at The George Washington University who used the first draft of this book for a graduate course on critical issues in distance education. Their feedback as students and "users" was invaluable. I would also like to thank many colleagues and reviewers for their comments and suggestions and in particular David Peal, for his critical review and helpful feedback. I would also like to thank Ruth Bennett for her review.

I would also like to thank many of my well-wishers who believed in me and were always there when I needed them: Larry Lipsitz, Charles Reigeluth, Fakhrul Ahsan, Saleha Begum, Anisur R. Khan, Maung Sein, Qamrul Islam, Syeda Munim, Ainul Abedin and A.K.M. Asaduzzaman.

Finally, and most important, I thank my wife Komar Khan and my sons Intisar Khan and Inshat Khan, my brothers Kamrul H. Khan, Manzurul H. Khan and Nazrul H. Khan, and my sisters Nasima Zaman and Akhtar Janhan Khanam, my cousins Mahmudul Alam and Masud Ul Alam, my childhood mentor Omar Ali, my niece Sabrina Zaman Choudhury and my nephews Murad, Shahid, Shuvo, Jamshed, Habib, Rakib, Ria, Shanta, Ruma, Shuma, Sajib and Sabuz for their continued support and encouragements. I would also like to thank all my nieces and nephews for their encouragement.

Chapter 1

Introduction

Introduction

Advances in information technology, coupled with the changes in society, are creating new paradigms for education and training. These massive changes have tremendous impact on our educational and training systems. Participants in this educational and training paradigm, require rich learning environments supported by well-designed resources (Reigeluth & Khan, 1994). They expect on-demand, anytime/anywhere high-quality instruction with good support services. To stay viable in this global competitive market, providers of education, and training must develop efficient and effective learning systems to meet the society's needs. Therefore, there is a tremendous demand for affordable, efficient, easily accessible, open, flexible, well-designed, learner-centered, distributed, and facilitated learning environments.

Rosenberg (2001) stated, "Internet technologies have fundamentally altered the technological and economic landscapes so radically that it is now possible to make quantum leaps in the use of technology for learning." Hall (2001) reports that "e-learning is the fastest-growing and most promising market in the education industry. According to WR Hambrecht & Co., e-learning is poised to explode, and the company anticipates the market to more than double in size each year through 2002."

The Primary Research Group conducted a survey of distance learning programs in higher education in 2002. The survey results show continued astounding growth in the higher education distance learning market. The mean annual enrollment growth rate for 2002 reported by the 75 college distance learning programs in the study was 41 percent, and 92 percent of programs sampled say that their enrollment growth rate has been either "very strong" or "fairly good." Not a single college in the survey experienced a decline in total distance learning course enrollment. In 2004 survey, the Primary Research Group found that college distance learning programs increased their revenues by a mean of 9.67 percent in 2003 (source: The Survey of Distance & Cyberlearning Programs in Higher Education, 2002 Edition and 2004 Edition).

Eduventures Inc., a Boston-based research firm, announced that it expects total enrollment in online education programs to hit the one million mark by 2005. The report, Online Distance Education Market Update: A Nascent Market Matures, found that the majority of U.S. colleges and universities offer some form of online education, helping the market grow more than 50 percent in 2002 to reach $3.7 billion. The report also estimates that the growth rate for online education will exceed 30 percent for "a number of years to come" (source: *Boston Business Journal*, http://boston.bizjournals.com/boston/stories/2004/03/08/daily22.html).

With the advent of the Internet and online learning methodologies and technologies, providers of education (K-12 and higher education) and training are creating e-learning materials to fulfill the demand. Online learning is becoming more and more accepted in workplace. Institutions are investing heavily in the development and deployment of online programs. Academic institutions, corporations, and government agencies worldwide are increasingly using the Internet and digital technologies to deliver instruction and training. At all levels of these institutions, individuals are being encouraged to participate in online learning activities. Many instructors and trainers are being asked by their institutions to convert their traditional face-to-face (f2f) courses to e-learning. Individuals involved in designing e-learning or converting f2f courses to online environments are faced with many challenges, such as: What is e-learning and how is it different from f2f learning? What does and does not work for e-learning? How does one measure e-learning success?

In this chapter, e-learning is discussed from the perspectives of open, flexible, and distributed learning environment and how its various learning features can be designed to address critical issues of e-learning environment. The following is an outline for the chapter:

- What is e-learning?
- Open, flexible, and distributed learning environment
- Traditional instruction and e-learning
- Learner-focused e-learning system
- Components and features of e-learning
- A framework for e-learning
- Review of e-learning features with the framework

What is E-Learning?

With the Internet's and digital technologies' rapid growth, the Web has become a powerful, global, interactive, dynamic, economic, and democratic medium of learning and teaching at a distance (Khan, 1997). The Internet provides an opportunity to develop learning-on-demand and learner-centered instruction and training. There are numerous names for online learning activities, including e-learning, Web-based learning (WBL), Web-based instruction (WBI), Web-based training (WBT), Internet-based training (IBT), distributed learning (DL), advanced distributed learning (ADL), distance learning, online learning (OL), mobile learning (or m-learning) or nomadic learning, remote learning, off-site learning, a-learning (anytime, anyplace, anywhere learning), and so on. In this book, the term e-learning is used to represent open, flexible, and distributed learning.

Designing and delivering instruction and training on the Internet requires thoughtful analysis and investigation, combined with an understanding of both the Internet's capabilities and resources and the ways in which instructional design principles can be applied to tap the Internet potential (Ritchie & Hoffman, 1997). Designing e-learning for open, flexible, and distributed learning environments is new to many of us. After reflecting on the factors that must be weighed in creating effective open, distributed, and flexible learning environments for learners worldwide, the following definition of e-learning is formulated in this book.

E-learning can be viewed as an innovative approach for delivering well-designed, learner-centered, interactive, and facilitated learning environment to anyone, anyplace, anytime by utilizing the attributes and resources of various digital technologies along with other forms of learning materials suited for open, flexible, and distributed learning environment.

The above definition of e-learning raises the question of how various attributes of e-learning methods and technologies can be utilized to create learning features appropriate for diverse learners in an open, flexible, and distributed environment.

Open, Flexible, and Distributed Learning Environment

A clear understanding of the open, flexible, and distributed nature of online learning environment will help us create meaningful e-learning. According to Calder and McCollum (1998), "The common definition of open learning is learning in your own time, pace, and place" (p. 13). Ellington (1997) notes that open and flexible learning allows learners to have some say in how, where, and when learning takes place. Therefore, in this book, I use the terms open and flexible to mean in your own time, pace, and place. Saltzberg and Polyson (1995) noted that distributed learning is not synonymous with distance learning, but they stress its close relationship with the idea of distributed resources:

Distributed learning is an instructional model that allows instructor, students, and content to be located in different, non-centralized locations so that instruction and learning occur independent of time and place . . . The distributed learning model can be used in combination with traditional classroom-based courses, with traditional distance learning courses, or it can be used to create wholly virtual classrooms. (p. 10)

The Internet supports open learning because it is device, platform, time, and place independent. It is designers who take advantage of the openness of the Internet to create learning environments that are flexible for learners. Therefore, openness is a technical matter; flexibility is a design matter. The Internet, by its very nature, distributes resources and information, making it the tool of choice for those interested in delivering instruction using the distributed learning model (Saltzberg & Polyson, 1995). Thus, the Internet supported by various digital technologies is well-suited for open, flexible, and distributed learning (Figure 1).

In reviewing the manuscript of this chapter, Janis Taylor (2004) of Clarke College provided the following suggestions to clarify the terms open, flexible, and distributed which I find useful:

It is hard to remember the definition of e-learning as open, flexible, and distributed just from the words. Consider a student user whose quote

Figure 1. Open, flexible, and distributed e-learning

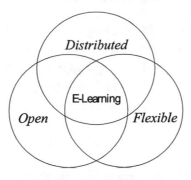

described her online education as open because she can sit out on her back deck supervising her children in the swimming pool while doing her homework. Now that's "open-air" and "open" learning. Listen to a young traveling businesswoman who says: "I take my college course, my instructor, and all of my fellow students with me on every business trip. With my laptop in my hotel room, I can view my teacher's demonstration, discuss it with my classmates in the Chat Room, and turn in my assignment by e-mail. Now that's a flexible college program!" Hear the instructor who says: "One of my pre-service teachers works in a chemical lab in Cleveland three states away, another is a court reporter three hours drive from me, another is a nurse in rural western Iowa far from any of our teacher's colleges, and another a physician who wants to keep his practice going until he makes the career change from medicine to teaching science. I, their teacher, am sitting in a small liberal arts college in eastern Iowa, a state badly needing to tap new people to come into the teaching profession. How could I get them all here to my campus if e-learning weren't distributed? E-learning distributes instructional material from me to them and the Internet distributes what they are learning back to their fellow students and to me."

Traditional Instruction and E-Learning

The design and format of open, flexible and distributed e-learning can be different from traditional classroom instruction. Traditional classrooms are space bound. Traditional instruction treats learning pretty much as a closed system, taking place within the confines of a given classroom, school, textbook, field trip, and so on. (Greg Kearsley, personal communication, January 27, 2000). Classroom-taught courses are not necessarily closed systems; many teachers assign students to do library based research papers, interview members of a professional community, engage in service-learning activities, and extend their learning initiatives far beyond the classroom itself. Unfortunately, many classes are bound by their four walls involving only the thoughts of the instructor, the textbook writer and occasional student comments. Classroom courses are also closed in the sense that they are limited to only those students who can physically come to the location (Janis Taylor, personal communication, May 24, 2004).

On the other hand, e-learning extends the boundaries of learning, so that learning can occur in the classrooms, from home and in the work place (Relan & Gillani, 1997). It is a flexible form of education, because it creates options for learners in terms of where and when they can learn (Krauth, 1998). A well-designed e-learning course allows learners to become actively involved in their learning processes. However, a poorly designed e-learning course can be just as rigid and dogmatic and non-interactive as a poorly taught face-to-face course. The scope of openness and flexibility in e-learning is dependent on how it is designed. "While having an open system has its appeal, it can make designing for it extremely difficult, because in an open system, the designer agrees to give up a certain amount of control to the user" (Jones & Farquhar, 1997, p. 240). The more open the learning environment, the more complex the planning, management, and evaluation of it (Land & Hannafin, 1996). For example, the instructor cannot monitor who helps the student on tests unless proctored.

Learner-Focused E-Learning System

A leading theorist of educational systems, B.H. Banathy (1991), makes a strong case for learning-focused educational and training systems where "the learner is the key entity and occupies the nucleus of the systems complex of education" (p. 96). For Banathy, "when learning is in focus, arrangements are made in the environment of the learner that communicate the learning task, and learning resources are made available to learners so that they can explore and master learning tasks" (p. 101). A distributed learning environment that can effectively support learning-on-demand must be designed by placing the learners at the center. In support of learner-centered approach, Moore (1998) states:

Our aim as faculty should be to focus our attention on making courses and other learning experiences that will best empower our students to learn, to learn fully, effectively, efficiently, and with rewarding satisfaction. It is the responsibility of our profession to study ways of maximizing the potential of our environments to support their learning and to minimize those elements in their environments that may impede it. (p. 4)

Success in an e-learning system involves a systematic process of planning, designing, evaluating, and implementing online learning environments where learning is actively fostered and supported. An e-learning system should not only be meaningful to learners, but it should also be meaningful to all stakeholder groups including instructors, support services staff, and the institution. For example, an e-learning system is more likely to be meaningful to learners when it is easily accessible, clearly organized, well written, authoritatively presented, learner-centered, affordable, efficient, flexible, and has a facilitated learning environment. When learners display a high level of participation and success in meeting a course's goals and objectives, this can make e-learning meaningful to instructors. In turn, when learners enjoy all available support services provided in the course without any interruptions, it makes support services staff happy as they strive to provide easy-to-use, reliable services. Finally, an e-learning system is meaningful to institutions when it has a sound return-on-investment (ROI), a moderate to high level of learners' satisfaction with both the quality of instruction and all support services, and a low drop-out rate (Morrison & Khan, 2003).

In the next section, various attributes of the Internet and other digital technologies are discussed in terms of how they can be used to create meaningful learning environments.

Components and Features of E-Learning

An e-learning program is discussed here in terms of various components and features that can be conducive to learning. Components are integral parts of an e-learning system. Features are characteristics of an e-learning program contributed by those components. Components, individually and jointly, can contribute to one or more features (Khan, 2001b). For example, e-mail is an asynchronous communication tool (component) that can be used by both students and instructors to interact on learning activities. Therefore, with appropriate instructional design strategies, e-mail can be integrated in an e-learning program to create an interactive feature between students and the instructors. Think about it this way. While traveling on an airplane, passengers can use Airfone to communicate with others on the ground. In this case, Airfone, is a component of the airplane system that allows passengers to establish a synchronous communication (feature). Likewise e-mail, mailing lists, newsgroups and conferencing tools (components), along with appropriate instructional design principles and strategies can contribute to a collaborative feature for students working on a group project. The Web site http://BooksToRead.com/wbt/component-feature.ppt hosts a PowerPoint slide presentation emphasizing the point made above.

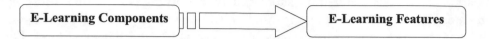

E-Learning Components

I have organized e-learning components into seven categories. As the e-learning methodologies and technologies continue to improve, components within the seven categories may need to be modified and new components may be available to be added. Components updates can be found at http://BooksToRead.com/wbt/component.htm). Please note that none of these components can create meaningful e-learning features without the proper integration of instructional design, which is included in the following list as one of the most important part of e-learning components.

1. **Instructional Design (ID)**
   (a) Learning and Instructional Theories
   (b) Instructional Strategies and Techniques

2. **Multimedia Component**
   (a) Text and Graphics
   (b) Audio Streaming (e.g., Real Audio)
   (c) Video Streaming (e.g., QuickTime)
   (d) Links (e.g., Hypertext links, Hypermedia links, 3-D links, imagemaps, etc.)

3. **Internet Tools**
   (a) Communications Tools
       (i) Asynchronous: E-mail, Listservs, Newsgroups, and so on.
       (ii) Synchronous: Text-based (e.g., Chat, IRC, MUDs, messaging, etc.) and audio-video conferencing tools.
   (b) Remote Access Tools (Login in to and transferring files from remote computers.)
       (i) Telnet, File Transfer Protocol (ftp), and so on.
   (c) Internet Navigation Tools (Access to databases and Web documents.)

        (i) Text-based browser, Graphical browser, VRML browser, and so on.

        (ii) Plug-ins

    (d) Search Tools

        (i) Search Engines

    (e) Other Tools

        (i) Counter Tool

4. **Computers and Storage Devices**

    (a) Computer platforms running Graphical User Interface (GUI) based operating systems such as Unix, Windows, Macintosh, Linux, and non-GUI based operating systems such as DOS. Mobile devices such as handheld personal digital assistants (PDAs) running Palm operating system, Pocket PC Windows, and other platforms.

    (b) Hard drives, CD ROMs, DVDs, and so on.

5. **Connections and Service Providers**

    (a) Modems

    (b) Dial-in (e.g., standard telephone line, ISDN, etc.) and dedicated (e.g., 56kbps, DSL, digital cable modem, T1, E1 lines, etc.) services (http://whatis.com/dsl.htm)

    (c) Mobile technology (e.g., connected wireless, wireless LAN, wireless WAN, wireless PAN or personal area network)

    (d) Application Service Providers (ASPs), Hosting Services Providers (HSPs), Gateway Service Providers, Internet Service Providers (ISPs), and so on.

6. **Authoring/Management Programs, Enterprise Resource Planning (ERP) Software, and Standards**

    (a) Scripting Languages (e.g., HTML - Hypertext Markup Language, VRML - Virtual Reality Modeling Language, XML – Extensible Markup Language, RSS - Rich Site Summary, is a text-based format, a type of XML http://www.faganfinder.com/search/rss.shtml#what, XSL - Extensible Style Sheet language, XHTML – Extensible Hypertext Markup Language, CSS - Cascading Style Sheets, WML-Wireless Markup language, Java, Java scripting, etc.).

(b) Learning Management System (LMS) and Learning Content Management System (LCMS)

(c) HTML Converters and Editors, and so on.

(d) Authoring Tools and Systems (easier to use than programming languages)

(e) Enterprise Application or Enterprise Resource Planning (ERP) Software in which e-learning solutions are integrated. (An article entitled "Integrating your Learning Management System with your Enterprise Resource Planning System" provides valuable information: http://www.thinq.com/pages/white_papers_pdf/ERP_%20Integration_0901.pdf)

(f) Interoperability, Accessibility, and Reusability Standards (http://www.adlnet.org/)

7.   **Server and Related Applications**

(a) HTTP servers, HTTPD software, and so on.

(b) Server Side Scripting Languages — JavaServer Pages (JSP), Active Server Pages (ASP), ColdFusion, Hypertext Preprocessor (PHP), Common Gateway Interface (CGI) — a way of interacting with the http or Web servers. CGI enables such things as image maps and fill-out forms to be run.

(c) Wireless Application Protocol (WAP) gateway — changes the binary coded request into an HTTP request and sends it to the Web server.

E-Learning Features

A well-designed e-learning program can provide numerous features conducive to learning. However, these features should be meaningfully integrated into an e-learning program to achieve its learning goals. The more components an e-learning program integrates, the more learning features it is able to offer. However, the effectiveness of e-learning features largely depends on how well they are incorporated into the design of the programs. The quality and effectiveness of an e-learning feature can be improved by addressing critical issues within the various dimensions of open, flexible, and distributed learning environment (discussed later in Table 3). The following are examples of some e-learning features: interactivity, authenticity, learner-control, convenience, self containment, ease of use, online support, course security, cost effectiveness, collaborative learning, formal, and informal environments, multiple expertise, online evaluation, online search, global accessiblity, cross-cultural interaction, non-

Table 1. Features and components associated with e-learning environments

E-Learning Features	E-Learning Components	Relationship to Open, Flexible, and Distributed Learning Environment
Ease of Use	A standard point and click navigation system. Common User Interface, Search Engines, Browsers, Hyperlinks, and so on.	A well-designed e-learning course with intuitive interfaces can anticipate learners' needs and satisfy the learners' natural curiosity to explore the unknown. This capability can greatly reduce students' frustration levels and facilitate a user-friendly learning environment. However, delays between a learner's mouse click and the response of the system can contribute to the frustration level of users. The hypermedia environment in an e-learning course allows learners to explore and discover resources which best suit their individual needs. While this type of environment facilitates learning, it should be noted that learners may lose focus on a topic due to the wide variety of sources that may be available on an e-learning course. Also, information may not always be accessed because of common problems related to servers such as connection refusal, no DNS entry, and so on (Khan, 2001b).
Interactivity	Internet tools, hyperlinks, browsers, servers, authoring programs, instructional design, and so on.	Interactivity in e-learning is one of the most important instructional activities. Engagement theory based on online learning emphasizes that students must be meaningfully engaged in learning activities through interaction with others and worthwhile tasks (Kearsley & Shneiderman, 1999). E-learning students can interact with each other, with instructors, and online resources. Instructors and experts may act as facilitators. They can provide support, feedback, and guidance via both synchronous and asynchronous communications. Asynchronous communication (i.e., e-mail. listservs, etc.) allows for time-independent interaction whereas synchronous communication (i.e., conferencing tools) allows for live interaction (Khan, 2001b).
Multiple Expertise	Internet and WWW	E-learning courses can use outside experts to guest lecturers from various fields from all over the world. Experiences and instruction that come directly from the sources and experts represented on the Internet can tremendously benefit learners.
Collaborative Learning	Internet tools, instructional design, and so on.	E-learning creates a medium of collaboration, conversation, discussions, exchange, and communication of ideas (Relan & Gillani, 1997). Collaboration allows learners to work and learn together to accomplish a common learning goal. In a collaborative environment, learners develop social, communication, critical thinking, leadership, negotiation, interpersonal, and cooperative skills by experiencing multiple perspectives of members of collaborative groups on any problems or issues.

Table 1. (continued)

E-Learning Features	E-Learning Components	Relationship to Open, Flexible, and Distributed Learning Environment
Authenticity	Internet and WWW, instructional design, and so on.	The conferencing and collaboration technologies of the Web bring learners into contact with authentic learning and apprenticing situations (Bonk & Reynolds, 1997). E-learning courses can be designed to promote authentic learning environments by addressing real world problems and issues relevant to the learner. The most significant aspect of the Web for education at all levels is that it dissolves the artificial wall between the classroom and the 'real world' (Kearsley, 1996).
Learner-Control	Internet tools, authoring programs, hyperlinks, instructional design, and so on.	The filtered environment of the Web allows students the choice to actively participate in discussion or simply observe in the background. E-learning puts students in control so they have a choice of content, time, feedback, and a wide range of media for expressing their understandings (Relan & Gillani, 1997). This facilitates student responsibility and initiative by promoting ownership of learning. The learner-control offered by e-learning is beneficial for the inquisitive student, but the risk of becoming lost in the Web and not fulfilling learner expectations can be a problem and will require strong instructional support (Duchastel, 1996).

discriminatory, and so on. As components of e-learning improve as a result of the advent of the Internet and online learning methods and technologies, existing e-learning features will improve and new features may be available to us. Several features that are contributed by e-learning components are presented in Table 1.

In designing e-learning environments using the features described above, we should explore (issues within the various dimensions) of open, flexible, and distributed learning environments (OFDLEs). In the next section, I present a framework that discusses issues of OFDLE. Also, later in this chapter, I review an e-learning feature for its effectiveness using issues within the various dimensions of OFDLE (Table 4).

A Framework for E-Learning

The seeds for the e-learning framework began germinating with the question, "What does it take to provide flexible learning environments for learners worldwide?" With this question in mind, since 1997, I have been communicating with learners, instructors, trainers, administrators, and technical and other support services staff involved in e-learning in both academic (K-12 and higher education) and corporate settings from all over the world. I researched critical e-learning issues discussed in professional discussion forums, and designed and taught online courses. I reviewed literature on e-learning. As the editor of *Web-Based Instruction* (1997), *Web-Based Training* (2001), and *Flexible Learning* (in press), I had the opportunity to work closely on e-learning issues with about two hundred authors from all over the world who contributed chapters in these books.

Through these activities, I have come to understand that e-learning represents a paradigm shift not only for learners, but also for instructors, trainers, administrators, technical, and other support services staff, and the institution. We (i.e., students, instructors, and staff) are accustomed to the structure of a traditional educational system where instructor-led, face-to-face classes are the learning environment. E-learning, on the other hand, is an innovative way of providing instruction to diverse learners in an environment where students, instructors, and support staff do not see each other. The format of such a learning environment is different from traditional classroom instruction. As indicated earlier, traditional classroom-based instruction takes place in a closed system (i.e., within the confines of a given classroom, school, textbook, or field trip), whereas e-learning takes place in an open system (i.e., it extends the boundaries of learning to an open and flexible space where learners decide where and when they want to learn). Learners in an open, flexible, and distributed learning environment need immediate attention and feedback on their work in order to continue their learning processes. We have to provide the best support systems for them so that they do not feel isolated and join the list of dropouts.

As we are accustomed to teaching or learning in a closed system, the openness of e-learning is new to us. In order to create effective environments for diverse learners, however, we need to jump out of our closed system learning design mentality. We need to change our mindset — that's the paradigm shift. In order to facilitate such a shift, and in response to the range of issues I saw in my research, I created the e-learning framework (Figure 2).

The purpose of this framework is to help you think through every aspect of what you are doing during the steps of the e-learning design process. Therefore, in this book, I am going to take each of the eight dimensions of this framework (for example, Chapter 1 deals with institutional dimension of the framework), and show what questions you should ask about each dimension as you design an e-learning segment either a lesson, a course or an entire program.

I found that numerous factors help to create a meaningful learning environment, and many of these factors are systemically interrelated and interdependent. A

Figure 2. The e-learning framework

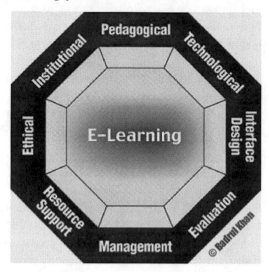

systemic understanding of these factors can help designers create meaningful learning environments. I clustered these factors into eight dimensions: institutional, management, technological, pedagogical, ethical, interface design, resource support, and evaluation (Table 2). Various issues within the eight dimensions of the framework were found to be useful in several studies that were conducted to review e-learning programs resources and tools (Barry, 2002; Chin & Kon, 2003; Dabbagh, Bannan-Ritland & Silc, 2001; El-Tigi & Khan, 2001; Gilbert, 2000; Goodear, 2001; Kao, Tousignant, & Wiebe, 2000; Khan & Smith, in press; Khan, Waddill & McDonald, 2001; Kuchi, Gardner & Tipton, 2003; Mello, 2002; Romiszowski, 2004; Singh, 2003; and Zhang, Khan, Gibbons & Ni, 2001).

Each dimension has several subdimensions (Table 3). Each subdimension consists of items or issues focused on a specific aspect of an e-learning environment. Throughout the book, the issues within each subdimension of the E-Learning Framework are presented as questions that course designers can ask themselves when planning e-learning. As you know each e-learning project is unique. I encourage you to identify as many issues (in the form of questions) as possible for your own e-learning project by using the framework. One way to identify critical issues is by putting each stakeholder group (such as learner, instructor, support staff, etc.) at the center of the framework and raising issues along the eight dimensions of the e-learning environment. This way you can identify many critical issues and answer questions that can help create a meaningful e-learning environment for your particular group. By repeating the same process for other stakeholder groups, you can generate a comprehensive list of issues for your e-learning project.

Table 2. Eight dimensions of e-learning framework

Dimensions of E-Learning	Descriptions
Institutional	The institutional dimension is concerned with issues of administrative affairs, academic affairs, and student services related to e-learning.
Management	The management of e-learning refers to the maintenance of learning environment and distribution of information.
Technological	The technological dimension of e-learning examines issues of technology infrastructure in e-learning environments. This includes infrastructure planning, hardware, and software.
Pedagogical	The pedagogical dimension of e-learning refers to teaching and learning. This dimension addresses issues concerning content analysis, audience analysis, goal analysis, media analysis, design approach, organization, and learning strategies.
Ethical	The ethical considerations of e-learning relate to social and political influence, cultural diversity, bias, geographical diversity, learner diversity, digital divide, etiquette, and the legal issues.
Interface design	The interface design refers to the overall look and feel of e-learning programs. Interface design dimension encompasses page and site design, content design, navigation, accessibility, and usability testing.
Resource support	The resource support dimension of the e-learning examines the online support and resources required to foster meaningful learning.
Evaluation	The evaluation for e-learning includes both assessment of learners and evaluation of the instruction and learning environment.

Table 3. Sub-dimensions of the e-learning framework

INSTITUTIONAL (Chapter 2)
Administrative Affairs
Academic Affairs
Student Services
MANAGEMENT (Chapter 3)
People, Process and Product (P3) Continuum
Management Team
Managing E-Learning Content Development
Managing E-Learning Environment
TECHNOLOGICAL (Chapter 4)
Infrastructure Planning
Hardware
Software

PEDAGOGICAL (Chapter 5)
Content Analysis
Audience Analysis
Goal Analysis
Design Approach
Instructional Strategies
Organization
Blending Strategies
ETHICAL (Chapter 6)
Social and Cultural Diversity
Bias and Political Issues
Geographical Diversity
Learner Diversity
Digital Divide
Etiquette
Legal Issues

INTERFACE DESIGN (Chapter 7)
Page and Site Design
Content Design
Navigation
Accessibility
Usability Testing
RESOURCE SUPPORT (Chapter 8)
Online Support
Resources
EVALUATION (Chapter 9)
Evaluation of Content Development Process
Evaluation of E-Learning Environment
Evaluation of E-Leaning at the Program
and Institutional Levels
Assessment of Learners

For example, is the course sensitive to students from different time-zones (e.g., are synchronous communications such as chat discussions are scheduled at reasonable times for all time zones represented)? This is an example of a question that e-learning designers can ask in the geographical diversity section of the ethical dimension. As we know scheduled chat discussions may not work for learners coming from different time zones. In the US, there are the six time zones. Therefore, e-learning course designers should be sensitive to diversity in geographical time zones (i.e., all courses where students can reasonably be expected to live in different time zones).

The purpose of raising many questions (which are included in Appendix A as e-learning quick checklist items) within each dimension (Table 3) is to help designers think through their projects thoroughly. Note that there might be other issues not included in the book, or not yet encountered. As more and more institutions offer e-learning worldwide, designers will become more knowledge-able about new issues within the eight dimensions of e-learning.

The e-learning framework can be applied to e-learning of any scope. This "scope" refers to a continuum defined by the extent to which instruction is delivered on the Internet and hence must be systematically planned for. The weight placed on any e-learning dimension or subdimension, or on any set of e-learning items, will vary with the scope of the instruction. This continuum is described below, with examples, to show the type and scope of e-learning activities and how their design relates to various dimensions of the framework.

At the "micro" end of the continuum, e-learning activities and information resources can be designed for face-to-face instruction in educational and training settings (e.g., blended learning activities). In the high-school physics classroom, for example, a teacher can use Shockwave simulations to support the cognitive work of analyzing data, visualizing concepts, and manipulating models. See, for example, the simulations available at http://www.explorescience.com. The teacher would have to design activities that provide context for and elaboration of this highly visual, Web-mediated simulation. In a traditional course, the e-learning framework's institutional and management dimensions will matter much less than the learning strategies section of the pedagogical dimension (Table 3), which provides guidelines for integrating the simulation into the curriculum.

Further along the continuum, more comprehensive design is required for the complete academic or training course, where content, activities, interaction, tutorials, project work, and assessment must all be delivered on the Internet. Petersons.com provides links to a large number of such courses that are exclusively or primarily distance-based. (The Petersons database can be searched at http://www.lifelonglearning.com). Additional dimensions of the e-learning framework will be useful in designing such courses.

Finally, at the "macro" end of the continuum, the e-learning framework can serve the design of complete distance-learning programs and virtual universities (Khan, 2001a), without a face-to-face component, such as continuing education programs for accountants or network engineers. Petersons.com, again, provides links to dozens of such programs as well as to institutions based on such programs. For example, designers of Web-based continuing education for accountants dispersed all around world would have to plan for every dimension of the e-learning framework in considerable detail. They would have to work with computer programmers, testing specialists, security professionals, subject-matter experts, and accountants' professional organizations. These designers would have to do everything from planning a secure registration system to considering cultural and language differences among accountants seeking continuing education credit.

As the scope of e-learning design expands, design projects change from one-person operations to complex team efforts. The e-learning framework can be used to ensure that no important factor is omitted from the design of e-learning, whatever its scope or complexity.

You might wonder: are all subdimensions within the eight dimensions necessary for e-learning? You might also think: there's a lot of questions here. Which ones do I need to address? Again, it depends on the scope of your e-learning initiative. To initiate an e-learning degree program, for example, it is critical to start with the institutional dimension of the e-learning framework and also investigate all issues relevant to your project in other dimensions. In this case, a comprehensive readiness assessment (see the readiness assessment section of institutional dimension in Chapter 2) should be conducted. However, to create a single e-learning lesson, some institutional subdimensions (such as admissions, financial aid, and others) may not be relevant.

Designing open, flexible, and distributed e-learning systems for globally diverse learners is challenging; however, as more and more institutions offer e-learning to students worldwide, we will become more knowledgeable about what works and what does not work. We should try to accommodate the needs of diverse learners by asking critical questions along the eight dimensions of the framework. The questions may vary based on each e-learning system. The more issues within the eight dimension of the framework we explore, the more meaningful and supportive a learning environment we can create. Given our specific contexts, we may not be able to address all issues within the eight dimensions of the framework, but we should address as many as we can.

Review of E-Learning Features with the Framework

As indicated earlier, all e-learning features must be designed to help students achieve their learning goals. An e-learning program consisting of well-designed instructional features can lead to its success. The eight dimensions of the e-learning framework can identify the critical issues of an e-learning environ-

Table 4. Review of an e-learning feature with the framework

E-Learning Feature	E-Learning Dimensions	Issues/Concerns
Ease of Use	*Institutional*	Are instructor/tutor and technical staff available during online orientation?
	Management	Does the course notify students about any changes in due dates or other course relevant matters such as server down?
	Technological	Are students taught how to join, participate in, and leave a mailing list?
	Pedagogical	Does the course provide a clear direction of description what learners should do at every stage of the course?
	Ethical	Does the course provide any guidance to learners on how to behave and post messages in online discussions so that their postings do not hurt others' feelings?
	Interface design	How quickly can users find answers to the most frequently asked questions on the course site?
	Resource support	Does the course provide clear guidelines to the learners on what support can and cannot be expected from a help line?
	Evaluation	If learners are disconnected during an online test, can they log back and start from where they left off?

ment, and provide guidance on addressing them. We can improve the effectiveness of an e-learning feature by answering the questions raised in the framework. For example, *ease of use* is one of most important features in an e-learning environment. In Table 4, an e-learning feature (i.e., ease of use) is reviewed for its effectiveness in a course from the perspective of each of the eight dimensions.

There are questions similar to those in Table 4 that can be used to review how a feature such as ease of use can be made a part of an e-learning program. All these questions in Table 4 cover the eight e-learning dimensions point to one critical element, is it really easy to use? Therefore, for each feature we should explore as many issues as possible within the eight dimensions of the e-learning environment.

Questions to Consider

Can you think of any e-learning components not included in the seven categories of components listed in the chapter?

Can you think of any e-learning features not included in the chapter?

Activity

1.   Using Internet search engines, locate at least one article that discusses various e-learning components listed in the seven categories. Discuss how they (i.e., components) contribute to various learning features.

References

Banathy, B.H. (1991). *Systems designs of education: A journey to create the future*. Englewood Cliffs, NJ: Educational Technology Publications.

Barry, B. (2002). ISD and the e-learning framework. Retrieved January 24, 2003, from *http://www.wit.ie/library/webct/isd.html*

Bonk, C.J. & Reynolds, T.H. (1997). Leraner-centered Web instruction for higher-order thinking, teamwork and apprenticeship. In B. H. Khan (Ed.), *Web-based instruction* (pp. 167-178). Englewood Cliffs, NJ: Educational Technology Publications.

Calder, J. & McCollum, A. (1998). *Open and flexible learning in vocational education and training*. London: Kogan Page.

Chin, K.L. & Kon, P.N. (2003). Key factors for a fully online e-learning mode: A delphi study. In G. Crisp, D. Thiele, I. Scholten, S. Barker, & J. Baron (Eds.), *Interact, Integrate, Impact: Proceedings of the 20th Annual Conference of the Australasian Society for Computers in Learning in Tertiary Education*. Adelaide, December 7-10.

Dabbagh, N.H., Bannan-Ritland, B., & Silc, K. (2000). Pedagogy and Web-based course authoring tools: Issues and implications. In B.H. Khan (Ed.), *Web-based training* (pp. 343-354). Englewood Cliffs, NJ: Educational Technology Publications.

Duchastel (1996). Design for Web-based learning. *Proceedings of the WebNet-96 World Conference of the Web Society*. San Francisco.

El-Tigi, M.A. & Khan, B.H. (2001). Web-based learning resources. In B.H. Khan (Ed.), *Web-based training* (pp. 59-72). Englewood Cliffs, NJ: Educational Technology Publications.

Ellington, H. (1995). Flexible learning, your flexible friend. In C. Bell, M. Bowden & A. Trott (Eds.), *Implementing flexible learning* (pp. 3-13). London: Kogan Page.

Gilbert, P.K. (2002). The virtual university an analysis of three advanced distributed leaning systems. Retrieved February 24, 2004, from *http://gseacademic.harvard.edu/~gilberpa/homepage/portfolio/research/pdf/edit611.pdf*

Goodear, L. (2001). *Cultural diversity and flexible learning.* Presentation of findings, 2001 Flexible Learning Leaders Professional Development Activity. South West Institute of TAFE. Australia. Retrieved February 24, 2004, from *http://www.flexiblelearning.net.au/leaders/events/pastevents/2001/statepres/papers/lyn-handout.pdf*

Hall, B. (2001). *E-learning: Building competitive advantage through people and technology.* A special section on e-learning by *Forbes* Magazine. Retrieved January 24, 2003, from *http://www.forbes.com/specialsections/elearning/*

Jones, M.G. & Farquhar, J.D. (1997). User interface design for Web-based instruction. In B.H. Khan (Ed.), *Web-based instruction* (pp. 239-244). Englewood Cliffs, NJ: Educational Technology Publications.

Kao, D., Tousignant, W. & Wiebe, N. (2000). A paradigm for selecting an institutional software. In D. Colton, J. Caouette, and B. Raggad (Eds.), *Proceedings ISECON 2000*, v 17 (Philadelphia): 207. AITP Foundation for Information Technology Education.

Kearsley, G. (1996). The World Wide Web: Global access to education. *Educational Technology Review,* Winter(5), 26-30.

Kearsley, G. & Shneiderman, B. (1999). Engagement theory: A framework for technology-based teaching and learning. (*http://home.sprynet.com/~gkearsley/engage.htm*)

Khan, B.H. (1997). Web-based instruction: What is it and why is it? In B.H. Khan (Ed.), *Web-based instruction* (pp. 5-18). Englewood Cliffs, NJ: Educational Technology Publications.

Khan, B.H. (2001a). Virtual U: A hub for excellence in education, training and learning resources. In B.H. Khan (Ed.), *Web-based training* (pp. 491-506). Englewood Cliffs, NJ: Educational Technology Publications.

Khan, B.H. (2001b). Web-based training: An introduction. In B.H. Khan (Ed.), *Web-based training* (pp. 5-12). Englewood Cliffs, NJ: Educational Technology Publications.

Khan, B.H. & Ealy, D. (2001). A framework for Web-based authoring systems. In B.H. Khan (Ed.), *Web-based training* (pp. 355-364). Englewood Cliffs, NJ: Educational Technology Publications.

Khan, B.H. & Smith, H.L. (in press). Student evaluation instrument for online programs. In B.H. Khan (Ed.), *Flexible learning.* Englewood Cliffs, NJ: Educational Technology Publications.

Khan, B.H., Waddill, D. & McDonald, J. (2001). Review of Web-based training sites. In B.H. Khan (Ed.), *Web-based training* (pp. 367-374). Englewood Cliffs, NJ: Educational Technology Publications.

Krauth, B. (1998). Distance learning: The instructional strategy of the decade. In G.P. Connick (Ed.). *The distance learner's guide*. Upper Saddler River, NJ: Prentice Hall.

Kuchi, R., Gardner, R. & Tipton, R. (2003). A learning framework for information literacy and library instruction programs at Rutgers University Libraries. Recommendations of the Learning Framework Study Group. Rutgers University Libraries.

Land, S.M. & Hannafin, M.J. (1997). Patterns of understanding with open-ended learning environments: A qualitative study. *Educational Technology Research and Development, 45*(2), 47-73.

Mello, R. (2002, June). 100 pounds of potatoes in a 25-pound sack: Stress, frustration, and learning in the virtual classroom. *Teaching With Technology Today, 8*(9). Retrieved February, 2004, from *http://www.elearningmag.com/elearning/article/articleDetail.jsp?id=2031*

Moore, M.G. (1998). Introduction. In C.C. Gibson (Ed.), *Distance learners in higher education*. Madison, WI: Atwood Publishing.

Morrison, J.L. & Khan, B.H. (2003). The global e-learning framework: An interview with Badrul Khan. *The Technology Source*. A Publication of the Michigan Virtual University. Retrieved May 18, 2003, from *http://ts.mivu.org/default.asp?show=article&id=1019#options*

Reigeluth, C.M. & Khan, B.H. (1994). Do instructional systems design (ISD) and educational systems design (ESD) really need each other? Paper presented at the *Annual Meeting of the Association for Educational Communications and Technology (AECT)*, Nashville, TN, February.

Relan, A. & Gillani, B.B. (1997). Web-based instruction and traditional classroom: Similarities and differences. In B.H. Khan (Ed.), *Web-based instruction* (pp. 41-46). Englewood Cliffs, NJ: Educational Technology Publications.

Ritchie, D.C. & Hoffman, B. (1997). Incorporating instructional design principles with the World Wide Web. In B.H. Khan (Ed.), *Web-based instruction* (pp. 135-138). Englewood Cliffs, NJ: Educational Technology Publications.

Romiszowski, A.J. (2004). How's the e-learning baby? Factors leading to success or failure of an educational technology innovation. *Educational Technology, 44*(1), 5-27.

Rosenberg, M.J. (2001). *E-Learning: Strategies for delivering knowledge in the digital age*. New York: McGraw-Hill.

Saltzbert, S. & Polyson, S. (1995). Distributed learning on the World Wide Web. *Syllabus, 9*(1), 10-12.

Singh, H. (2003). Building effective blended learning programs. *Educational Technology, 44*(1), 5-27.

Zhang, J., Khan, B.H., Gibbons, A.S., & Ni, Y. (2001). Review of Web-based assessment tools. In B.H. Khan (Ed.), *Web-based training* (pp. 137-146). Englewood Cliffs, NJ: Educational Technology Publications.

Chapter 2

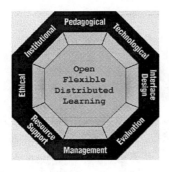

Institutional Issues

Institutions should develop comprehensive strategic and business plans for successful e-learning initiatives. Political factors often have significant impact upon the success of e-learning (Berge, 2001). Institutional funding and resources for delivering and maintaining e-learning are critical. Therefore, e-learning strategies must be aligned with and fully supported by the institutions' missions and strategic plans. E-learning initiatives require orchestration of personnel with diverse skills sets (Belanger & Jordan, 2000).

Institutions offering e-learning should consider online students as the consumers of education and training in a competitive market. Since more and more institutions offer e-learning programs, learners have more options to compare quality, services, price, and convenience of education providers. It should not be surprising that distance learners demand far more services than traditional campus-based students. Therefore, institutions should be ready to provide high quality education and training with the best learning resources and support services.

Some institutional issues (e.g., admissions, financial aid) may not be relevant to e-learning projects in other settings such as corporate training. However, many of the issues discussed in the institutional dimension can provide valuable insights into e-learning initiatives in K12 and corporate education arena.

The institutional dimension of e-learning is concerned with issues of administrative affairs, academic affairs, and student services related to e-learning. The following is an outline of this chapter:

- Administrative affairs
- Academic affairs
- Student services

Administrative Affairs

E-learning administrative affairs encompass issues related to needs assessment, readiness assessment, organization and change (diffusion, adoption, and implementation of innovation), budgeting and return on investment, partnerships with other institutions, program and course information catalog, marketing and recruitment, admissions, financial aid, academic calendar and course schedule, tuition and fees, registration and payment, information technology services, instructional design and media services, graduation, transcripts, and grades.

Needs Assessment

Needs analysis can help institutions to match the needs of their target audience with the e-learning courses and programs they plan to market. Any institution venturing into e-learning should conduct a needs assessment survey to find out its expected customers' (i.e., learners') willingness to enroll in its e-learning courses. Needs analysis will help institutions analyze the short-term and long-term needs for their e-learning initiatives, and in turn will be instrumental in developing their e-learning strategies. Needs analysis can also provide information about the technological and other support services needed for their e-learning initiatives. Through a comprehensive needs assessment process, an institution can establish its e-learning goals (discussed in goals analysis section of Chapter 3).

Readiness Assessment

Readiness assessment helps to review the comprehensive readiness status of an institution's e-learning initiative and it also points to critical factors that should be considered in order to get ready for e-learning. Open, flexible and distributed nature of e-learning environment requires that we review the status of readiness in all possible domains. Chapnick (2001) discusses eight types of readiness: psychological, sociological, environmental, human resources, financial, technological, equipment, and content. Welsch (2002) discusses three basic types of readiness: financial, structural and cultural. In this book, financial, infrastructure, cultural, and content readiness are discussed.

Financial Readiness

Any e-learning initiative requires financial support by the institution. An institution should analyze whether its e-learning initiative is worth the investment. One of the most critical financial readiness factors is the consideration of the long-term budget for e-learning initiatives.

Infrastructure Readiness

An e-learning initiative requires a well-planned technological infrastructure supported by adequate resources. Infrastructure readiness includes human resource readiness, equipment readiness, and technological skill readiness.

Cultural Readiness

Self-directed e-learning can face challenges where face-to-face learning has strong tradition. Therefore, it is important for institutions to assess their institutional cultures by exploring learners' learning preferences, instructors' teaching preferences, and the existing learning culture. Cultural readiness includes psychological, sociological, and environmental readiness. Chapnick (2000) noted that the psychological readiness factor considers the individual's state of mind as it impacts the outcome of the e-learning initiative, the sociological readiness factor considers the interpersonal aspects of the environment in which the program will be implemented, and the environmental readiness factor considers the large-scale forces operating on the stakeholders both inside and outside the organization (http://www.learningcircuits.org/nov2000/chapnick.html).

Content Readiness

E-learning contents are organized and based on the goal(s) of an e-learning project. E-learning contents may include various multimedia components including text, graphic, audio, video, and animation. A complete assessment of these components is critical to the design of e-learning. It takes a lot of time and effort to create audio, video, and animations for an e-learning project. In some e-learning projects, text may be ready, but multimedia components may be under construction. This incompatibility delays the development of the finished project. Please note that existing contents may need to be modified during the e-learning development process, original text in particular may need to be rewritten, reorganized, and chunked.

Organization and Change (Diffusion, Adoption, and Implementation of Innovation)

Advances in technology are continually changing the way we learn, live, work, and think. With the blessing of the Internet, now we can engage in learning activities without having to come to face-to-face classrooms. Technology allows us to collaborate on projects with people we never met. The impact of technology is dramatically changing the nature of organizations and our ideas about learning and knowledge (Kearsley & Marquardt, 2001). In this technology-based society, institutions should take advantage of the Internet and various digital technologies to improve learning environments and become e-learning organizations. "E-learning involves a change of paradigm — to some degree, a change in how you deal with knowledge and information in your organization" (Andy Snider in Schelin, 2001).

In this new paradigm of learning, institutions must develop a coordinated vision for technological change that can effectively guide the e-learning process (Rossner & Stockley, 1997). A vision of technological change should begin with the straightforward realization that the institution has a stated mission and a related set of shared values concerning the teaching, research, and administrative functions that must be served by the new technology (Gilbert, 1996, cited in Rossner & Stockley, 1997).

To become e-learning organizations, institutions (with the hierarchical organizational structures) may need to fundamentally alter the way their organizations are structured. The organizational structure of e-learning organizations should be flexible enough to accommodate the ever-changing needs of learners. Kearsley and Marquardt (2001) notes that the new building blocks of e-learning organizations are their "infostructures" — organizational structures built on information, learning and technology. Kearsley and Marquardt (2001) emphasize that institutions will need to continuously transform themselves into learning organizations with new infostructures in order to become places where groups and individuals continuously engage in new learning processes. Therefore, changes in how organizations are structured become a major step toward initiating e-learning.

E-learning is an innovative way of providing learning to diverse learners in an open, flexible, and distributed environment. This type of learning environment is new to many of us, therefore, institutions should make special effort to get greater support and acceptance from all stakeholder groups including learners, instructors, support services staff, and community members. For corporate settings, input and influence from stakeholders such as employees and customers will be very useful. Stakeholders will greatly participate in e-learning if they are well-informed about the benefits of e-learning in their professional and personal lives. Institutions should use diffusion, adoption, and implementation strategies for e-learning initiatives.

Diffusion and Adoption

Diffusion involves spreading the word about the e-learning initiative and thus can contribute to the adoption of e-learning. It is important for the institution to form a change team for e-learning with individuals who are technology literate. The team can help people to understand the changes brought on by new technologies and show them the value and benefits of adopting new technologies. These "change agents" can activate the e-learning diffusion process and facilitate the adoption of e-learning among all stakeholders groups including faculty members, staff, learners, community members, and so on. All stakeholders group should be informed about how emerging technologies are changing the way we teach, train, learn, and do businesses. They should understand the value and benefits of emerging technologies in e-learning. Change team should thus help stakeholders to adapt to change and accept e-learning as a viable medium of instruction, training, and learning.

The diffusion process for e-learning can start with the use of surveys and interviews to identify those faculty members, training and support staff within the institution to whom their peers looked for direction in the use of technology (Jennings & Dirksen, 1997). These "opinion leaders" are respected members of the group who others look to for leadership. "Promotional efforts by an opinion leader is an effective and concrete promotional tool for facilitating change" (Jennings & Dirksen, 1997, p. 112). Institutions should consider giving these leaders release time or other incentives to train others to adopt e-learning.

Institutions should explore issues that may serve as obstacles to adoption of e-learning. Passmore (2000) analyzes some of the impediments that limit participation of university faculty in distance education. He discusses three impediments to Web-based course delivery that are faced by university faculty members: (1) limited access to and experience with resources for Web-based design, development, and delivery, (2) uncertainties about status of intellectual property created for Web-based courses, and (3) lack of a reward system tied to innovation in instruction.

Implementation

After the adoption process comes implementation. The implementation process benefits from an increased adoption of e-learning by stakeholder groups and the establishment of an effective and reliable online infrastructure for learning.

Ely (1999) identifies eight common conditions for successful implementation efforts of educational technology innovations: (1) dissatisfaction with the status quo, (2) knowledge and skills required by the user of innovation, (3) resources required to make implementation work, (4) time required for implementators to acquire knowledge and skills, plan for use, and adapt, integrate and reflect upon what they are doing, (5) rewards or incentives for users of the innovation, (6) shared decision-making and communications among all parties involved, (7) endorsement and continuing support for implementation of the innovation by the important individual(s) or groups within the organization, and (8) leadership of the executive officer of the organization and project leadership involved in day-to-day activities of the innovation. Institutions interested in e-learning should review these eight common conditions to plan their e-learning implementations.

Budgeting and Return on Investment

An institution should have a comprehensive budget plan for e-learning. There are three types of budgeting related to e-learning. Boettcher (1999) notes that the first budget is for the initial design and development of a program; the second budget is for the marketing and delivery of a program, the third budget is for the ongoing maintenance of a program. Institutions must have adequate funds for the three budgeting types in order to operate successful e-learning programs.

E-learning projects should be both instructionally and financially successful. A successful e-learning program should conduct a return on investment (ROI) study and develop effective marketing strategies. Gustafson and Schrum (2001) recommend that institutions do costing and ROI analysis before, during, and after the implementation of e-learning projects. Return on investment in e-learning involves comparing the costs of e-learning to its benefits. Therefore, the formula for ROI is:

$$\text{ROI (\%)} = \frac{\text{Benefits} - \text{Total Costs}}{\text{Total Costs}} \times 100$$

Gustafson and Schrum (2001) suggest that the following should be taken into consideration for full costing in e-learning:

- Instructor compensation (salaries and benefits)
- Travel and per diem for instructors
- Support staff compensation (salaries and benefits)
- Initial program development
- Program revision
- Equipment and software acquisition
- Maintenance and upgrade of equipment and software
- Consumable program materials
- Delivery (e.g., satellite or internet charges)
- Other operating costs (telephone, postage, supplies, publicity, etc.)
- Facilities (optional)

There are two types of benefits in e-learning: (1) tangible or "hard" benefits and (2) intangible or "soft" benefits.

Tangible or "hard" benefits can be converted to a monetary or dollar value. For example, an institution can save travel and per diem for instructors if it offers courses on the Internet instead of flying (or transporting) them to remote sites. Intangible or "soft" benefits are difficult to convert to dollar figures. For example, improved cross-cultural communication skills by students from taking an e-learning course is difficult to convert to dollar value.

Partnership with Other Institutions

The capabilities of the Internet and digital technologies to support online learning are increasingly attracting more and more institutions in the e-learning arena. This is good for the e-learning field. Now, customers (i.e., learners) have more options to choose quality e-learning courses from institutions around the world. Only well-designed e-learning programs will attract customers. Two or more institutions with similar academic status can establish partnership for e-learning programs. They can market each other's e-learning courses to their students. It is a win-win situation for all parties involved. Since it takes a lot of efforts and time to design meaningful e-learning, it might be better for some institutions to offer joint programs. This type of partnership can be mutually beneficial for all partner institutions. For example, University of British Columbia in Canada has a partnership with the University of Queensland and The University of Melbourne in agricultural sciences (Bates, 2001).

Program and Course Information Catalogue

Institution should provide most up-to-date and accurate information about their courses and programs, academic calendar, schedule, and tuition and fees.

General Information

Institutions should provide information about degree requirement, course format, accreditation, and credit transfer relevant to all e-learning offerings.

Academic Calendar and Course Schedule

Institution should provide a complete listing of courses, duration (e.g., four-week, eight-week, 12-week, 16-week, etc.), start and end dates, and so on. All online synchronous activities' schedule should also be provided.

Tuition and Fees

Institutions offering e-learning courses should provide updated information about tuition and fees for their e-learning courses. They should also clearly indicate if any future increases in tuition and fees for any of their courses. Since e-learning market is increasingly competitive, reasonable tuition and fees are highly desirable by distance learners worldwide. Accurate information about tuition and fees is always helpful for distance learners as they have to make decisions for selecting online courses and programs. It is more likely that learners taking e-learning courses from partner institutions (see Partnership with Other Institutions section) will expect similar tuition and fees for all their courses toward a degree. Partner institutions may consider one umbrella tuition structure for their e-learning courses.

Marketing and Recruitment

The emergence of the Internet as a viable medium for distributed learning has attracted both academic and non-academic institutions into e-learning arena. These institutions see a great return on investment in e-learning. As a result, an increasing number of these institutions are currently either offering or in the process of offering e-learning courses/programs. Now, learners have more options to choose from a variety of e-learning courses or programs from all over the world that best suit their needs. This is good for learners, but makes the e-learning market very competitive. In this e-learning market, non-academic institutions or vendors often compete with academic institutions with their e-learning offerings.

It should be noted that only well designed e-learning courses with quality contents and good support services will attract learners (i.e., customers). In this competitive global economy, institutions must find ways to position their e-learning offerings most effectively to attract and retain a "critical mass" of learners (Lavenburg, 2001). For example, institutions can provide learners' testimonials on how well the courses and programs are designed for learning. Teikyo Post University provides testimonials from the students about their online learning experience (http://www.tpuonline.com/Testimonials/index.html).

Ongoing market research on e-learners (i.e., clients) can provide institutions with comparative advantage over others in their e-learning offerings. Market researchers and recruiters (or salespersons) should be part of the overall e-learning marketing initiative. The scope of this marketing operation may depend on institutions' e-learning policies and types of their clients (i.e., learners). One of the important marketing strategies is to make accurate and updated information about their e-learning offerings known to as many potential learners as possible. This can be accomplished by registering e-learning sites with search engines, banner advertising, postings in listservs, brand strategy (e.g., brand name), endorsement by credible people and institutions, and so on. Effective marketing will help institutions to attract and recruit students for their courses and programs.

Attrition or dropout rate becomes an important criterion for recruitment efforts. Higher dropout rate may pose problem in recruitment. Potential students will be more interested in programs whose students successfully complete courses. E-learning institutions should analyze the causes of dropout on a regular basis.

In their study of 118 distance education students, Morgan and Tam (1999) found several barriers: situational (e.g., lack of free time, change in circumstances, took more time than expected, study not related to job, etc.), dispositional (e.g., personal study problems, unclear goals, time management problems, etc.), institutional (e.g., problem with course schedule and pacing, learning materials arrive late, insufficient feedback on assignments, insufficient/unsatisfactory communication with academics, course focus and expectations not clear, missed contact with other students, inflexible course structure, problems getting academics to call back, course content outdated, confusing changes to course, etc.) and epistemological (e.g., difficult content, mismatch in assessment requirement, course focus lacked personal relevance or interest, lacked prerequisite knowledge, etc.).

Institutions offering e-learning courses should pay attention to these barriers in order to improve retention. For example, a barrier such as "insufficient feedback on assignments" definitely contributes to dropout rate. Kearsley (2002) states, "When students don't get enough feedback in an online course, they tend to drop out, producing completion rates of less than 50 percent" (p. 42). Therefore, institutions should periodically collect and analyze data on why students drop out of online courses, and find ways to minimize the attrition rate by systemically addressing factors contributing to the barriers.

Admissions

Prompt and reliable services from the admission office are very critical in e-learning. Admissions offices should be efficient and user friendly from the perspectives of human relations and technical capabilities. Distance learners would greatly appreciate the fast and reliable services from admissions office. Institutions offering e-learning should consider installing secure and reliable online systems to accept students' application forms for admissions.

Financial Aid

For e-learning, financial aid services should be organized to provide the best support services for distance learners by using both technological and human support services. Students should be able to speak to financial aid advisers to get guidance on their educational finances. Financial aid, student loan and scholarship information should be available online. Institutions can provide online financial aid workshops to assist students with both financial aid forms and other scholarship opportunities. University of Minnesota is the first institution in the United States to adopt Web-based financial aid program, a paperless system for financial aid (Mary Jane Smetanka, *Star Tribune*, April 26, 2001). Dominican University provides online student loan counseling services for students by providing links to site where students can complete a student loan entrance counseling session prior to obtaining a federal Stafford student loan (http://mapping-your-future.org/).

Registration and Payment

Institutions offering e-learning should have a secure and reliable system to handle all financial transactions. As the Internet and e-commerce technology continue to improve, procedures for automated registration, and online financial transactions are becoming an integral part of e-learning institutions.

Information Technology Services

Information Technology Services (ITS) is an important component of an e-learning initiative. ITS includes network and computing support services for faculty, staff, and learners. These services may include managing application software and servers for courses, providing e-mail accounts, disk space for Web pages, technical support for students, faculty members, and staff (also discussed in Chapter 4). ITS should be able to provide ongoing support services for e-learning technologies. Whenever appropriate, ITS should work with e-learning staff to choose learning management system (LMS) and learning content management system (LCMS) for courses. According to Paul Stacey (2001), "An LMS is for training managers, instructors, and administrators providing primary management of learners. An LCMS is for content developers, instructional designers, and learning managers providing primary management of learning content."

LMS and LCMS are discussed in detail in Chapter 4.

Instructional Design and Media Services

Instructional design and media services department can provide key services during the design and development processes of e-learning. Designing and delivering e-learning requires thoughtful analysis and investigation of how to use the Internet in concert with instructional design principles while considering the eight dimensions of open, flexible, and distributed e-learning environment. Usually, individuals with skills in instructional design work with content or subject matter experts to design the blueprint for e-learning. Programmers, graphic artists, and multimedia designers work together as team for the development or creation of e-learning materials by following the blueprint.

Graduation, Transcripts, and Grades

Graduation

The graduation ceremony is the celebration of successful completion of great endeavors. It is not easy to complete a certificate or degree program completely online. It takes a lot of patience and commitment. At the end of this successful completion, these students should be awarded their efforts. In some institutions, online students can attend a graduation ceremony at the campus. In virtual universities, online students can join cyber graduation. For example, Jones International University (JIU), the first fully-online accredited university invited

Erin Brockovich (subject of the critically acclaimed, Academy Award-winning film "Erin Brockovich") to deliver the commencement address at JIU's Cyber Graduation on May 22, 2001 (*e-learning magazine*, June 2001, p. 12).

Institutions should provide graduation information, including application forms, graduation ceremonies, and so on to distance learners online. Institutions should also provide graduation advising for distance learners so that they complete all requirements in order to graduate.

Transcripts and Grades

An e-learning institution should develop an online system that allows distant students to securely access their up-to-date academic records and grades. Just-in-time accessibility to transcripts and grades are some of the most important services that institutions can provide to remote learners. Institutions can encourage faculty to submit their students' grades on time using an online system so that students can get the most up-to-date grades and records. Faculty should be encouraged to use the online system (if available) to change grades as needed.

Academic Affairs

Academic affairs can encompass accreditation, policies, instructional quality, faculty and staff support, workload, class size compensation, intellectual property rights and so on.

Accreditation

The assurance of quality education in e-learning is critical. For online courses, learners will demand high quality e-learning environments supported by well-designed learning resources from accredited institutions. Therefore, institutions offering e-learning should receive appropriate accreditation (whenever needed) from a recognized accrediting agency for their e-learning programs. Students definitely want to obtain credits from online courses that are respected and accredited. It is important for communities around the world to make sure that accreditation standards used by accrediting agencies are based on the assurance that the distance learner receives the best-quality learning environment with good support services and resources. In the United States, for example, Jones International University received formal institutional accreditation from the Higher Learning Commission, a member of the North Central Association, an accrediting body for institutions of higher education in the United States (http://www.ncahigherlearningcommission.org/resources/distance-learning/).

Policies

Institutions should develop their e-learning policies and guidelines. These institutional e-learning policies and guidelines must be communicated to all stakeholders groups including instructors, students, and support staff. East Carolina University (http://www.ecu.edu/webdev/policy.html) has developed policies and guidelines for the content and appearance of documents and other subject matter contained on all Web pages.

Instructional Quality

Instructional quality in e-learning is completely dependent on how well the learning environment is designed and managed, and how dedicated and involved are the instructional and support staff. A dedicated instructional and support staff can help create meaningful learning environment for learners.

An online course demands more time and effort from the instructor. It is advisable to limit the number of students per instructor (see section on *Workload, Class Size and Compensation)*. This should be a manageable number so that learning is truly fostered and supported by the instructor. To provide the best, most meaningful learning environments, instructors should have enough time to interact with students in their learning processes. An institution which uses several teaching assistants under a supervisory instructor (I call it "factory model") and offer online courses to unlimited number of students may not serve learners the same way when one instructor teaching one online course with a limited number of students.)

Faculty and Staff Support

Faculty members and staff involved in e-learning should receive proper training and resources to be effective in teaching and supporting student's learning. Price (1999) notes that faculty training typically focuses on technical matters such as using authoring software rather than instructional design. He believes that this insufficient training leads to the production of mediocre courses. Bischoff (2000) notes that online education's effectiveness lies in large part with the facilitators (i.e., online instructors) who must maintain visibility, give regular feedback, provide high-quality materials, and remove obstacles to student retention.

Institutions should provide adequate support and incentives to teach e-learning courses. These supports should also include financial support for faculty members and staff to conduct research, attend conferences, and present papers in professional meetings.

To understand the complexity of e-learning environment and what it takes to provide the best learning experience to learners, I recommend that administrators and instructors either teach or take one online course. Many of the faculty members who are currently teaching online courses might not have taken online courses during their college lives (since online course offerings were not available then). Therefore, it seems necessary for instructors who are planning to teach online to consider taking at least one online course plus some ongoing faculty development training on issues of e-learning. Instructors who took online courses as part of their college curriculum can be exempt from taking online courses. Each institution can make its own decisions about the skills needed to teach online. Stan Trollip posted the following message on the Instructional Technology Forum listserv on March 22, 2001 about online instructors taking online courses:

Ensure that the instructor/facilitator is well prepared. This is critical, as so few people have the skills (also, good stand-up teachers do not necessarily make good online instructors). At Capella, someone who wants to be an instructor must go through an online course about teaching online. It helps to see things from the other side. Then the first teaching experience is done in tandem with an experienced instructor. We have found that good facilitation makes a big difference.

University of Maryland University College provides online workshops for faculty. Workshop topics include (not limited to) such as academic policies that can affect instructors, understanding and working with students with disabilities, focus and strategies for assessing online and Web-enhanced courses, fair use, faculty ownership, and the digital millennium copyright act, and handling difficult students in online courses (http://umuc.edu/facdev/workshops.html#Online).

In online learning, an instructor must have command of the subject matter and the ability to convey knowledge, and should be able to play roles of a facilitator, mentor and coach. Aggarwal (2000) states:

In Web-based education, the instructor's role is that of a facilitator, mentor, and coach. As a facilitator, the instructor needs to know how to facilitate discussions in small groups, keep students task-oriented, and move them toward some sort of consensus. In case of dominance by some group members, the instructor needs to intervene and encourage input from non-participating members. As a mentor and a coach, the instructor will have to advise students on their progress, provide one-on-one counseling and offer prompt and constructive feedback. (p. 13)

Workload, Class Size, and Compensation

Designing and teaching online courses demand more time and effort from faculty members. Romiszowski and Chang (2001) report that courses delivered by computer-mediated communication (CMC) almost always involve significantly more instructor time than conventionally delivered courses. After conducting a study of the two methods of teaching, Robert H. Jackson of University of Tennessee made a statement which should not be surprising to anyone involved in e-learning, "many instructors say teaching an online course is much more time-consuming than teaching a traditional class." Robert H. Jackson (assistant dean) and Cyndi Wilson Porter (universe online director) provided the following rationale why online education takes more time (excerpted from an article titled "Ain't Got Time to Teach," *New York Times*, January 22, 2001):

Jackson says online education forces instructors to spend more time both preparing and teaching a course. Porter says the fact that many instructors must spend time learning how to use the tools needed for online education is another factor in their reluctance. She also contends that online education affords those students who would not interact with an instructor in a classroom setting an opportunity to interact at will. This results in a greater time burden for the instructor without increasing the student's time burden, Porter claims.

In her article entitled "Online vs. Traditional Degrees," Jennifer M. Sakurai of *e-learning magazine* cited comments from Leonard Presby, a professor at William Paterson University in NJ who teaches online:

Presby feels that the student time commitment might be equivalent between the two environments, but the faculty member's commitment is definitely higher. "Professors are basically available 24 hours a day," he says. "I have to constantly check my e-mail because the rule of thumb is that students get a response within 24 hours during the week and 48 hours over the weekend. If an instructor used to spend five hours a week in one-to-one contact with his students, I can see that doubling in an online environment." (e-learning magazine, August/September 2002, p. 30)

In considering the greater demands placed on instructors to design, develop, and implement e-learning courses, Williams and Peters (1997) noted that a faculty member in higher education might ask the following questions: How many journal articles could be published in the same amount of time? What are my priorities, teaching and learning or keeping my position? If proper incentives are not given, non-tenured faculty members may not participate in e-learning because it demands more time and effort. Rather, they may choose to get involved with projects that are counted toward their promotion and tenure. Therefore, academic institutions should establish a system of rewarding faculty members for undertaking such challenging tasks. For corporate and K12 settings, institutions should consider introducing some innovative incentives for trainers, instructors, and staff for their participation in e-learning.

An example of faculty incentives may include, a faculty member can be awarded same credit for the development of an e-learning course as he or she would receive for the publication of an article in a professional journal or magazine. "Institutions must recognize distance learning development as scholarship that is suitable for tenure review" (Accetta, 2001). Institutions should develop clear policies about faculty workload, compensation, and intellectual property rights, which are some of the most discussed issues in many online discussion forums. *The Chronicle of Higher Education* hosted a live colloquy on April 27, 2000 on the issue of "Technology and Tenure." Participants discussed the question: How can colleges fairly evaluate faculty members who want their digital scholarship and teaching to count when it comes to decisions on tenure, promotions, and raises? Alan Rea of Western Michigan University (WMU) reported on this issue:

We have recently negotiated a new faculty contract. Part of this contract calls for a significant revision in the way WMU compensates (tenure, promotion, course releases, etc.) faculty who develop and deliver "electronically-purveyed instruction." While the original Article 30 deals more with courses delivered via videotapes, CODEC, and other media, a committee representing both faculty and administration has been formed to revisit and suggest revisions to Article 30 (http://www.wmich.edu/aaup/ 99Art30.htm) as part of the new contract.

Class size is an important academic issue for e-learning. Keeping an e-learning class to a manageable size is very critical because an online course demands more time and efforts than a traditional face-to-face course. Learners participate in more discussions in e-learning environment than traditional classes thus requiring more time from teaching staff. Greg Kearsley, who has experience in teaching online courses in several academic institutions, says that the idea that online courses can have an almost infinite number of students is probably the biggest myth of all. He notes that most online courses need to have a small class size (no more than 15-20 for a given instructor), due to usual high level of interaction that occurs. According to Kearsley (2002), "Online teaching is much more reflective and demanding than most forms of traditional classroom instruction. You have to be willing to spend a good deal of time interacting with students and explaining details of the subject over and over again" (p. 41).

In inquiring about the recommended class size for e-learning, I came across several discussions on the class size issues on several listservs. Steve Collins did a survey of higher education faculty in 2000 that he posted on the eModerators listserve. According to Collins, the consensus regarding online class size is 20 maximum and five minimum; with many indicating 15 as a preferred maximum. Most who responded indicated that they were convinced that quality of instruction in a class decreased as the number of students rose above the indicated maximum, and that this especially applied to amount of interactivity in a course. Catherine Schifter conducted a national study, published in the *Online Journal of Distance Learning Administration*, spring 2000, asking about models and practices regarding distance education courses. She noted, "There is also some evidence that, if the intent is an interactive course, more than 20 students often results in fewer students participating — which also happens to be the case in f2f courses" (Excerpt from her posting on the [eModerators] listserv, Subject: RE: [eModerators] Class Size, Date: Wed., 11 July 2001).

Intellectual Property Rights

Institutions should provide clear information about intellectual property rights. Any uncertainties about the status of intellectual property can create confusion among faculty.

On university campuses, one of the most controversial issues today relates to ownership of copyright for course materials developed by a faculty member on salary at the institution, and subsequent use of those materials by a different faculty member at a different campus of the institution. Who owns the course? What happens when the faculty members who developed online course leave the institution; can they take the courses with them? Likewise, does the institution have complete freedom to package, license, and sell instructors' work? "The real need is for an institution to have a clear statement of its policy and a mechanism to ensure that the issue of ownership is addressed as early as possible in the development process" (Twigg, 2000). University of Maryland University College established the Center for Intellectual Property and Copyright (IP) that provides resources and information in the areas of intellectual property, copyright, and the emerging digital environment. The center provides workshops, online training, and electronic and print publications, and it provides continuous updates on legislative developments at the local, state, national, and international level (http://umuc.edu/distance/odell/cip/).

Student Services

Students taking e-learning courses should receive appropriate academic and support services as those taking face-to-face courses. It is also important to note that support services for distance learners may be different than traditional face-to-face students as their needs are unique. Student support services for e-learning should be effective and comprehensive. According to Connick (1998), "Students often assume that an institution offering courses or programs at a distance will also provide all of the essential services at a distance" (p. 26). These services include pre-enrollment services, orientation, advising, counseling, learning skills development services, services for students with disabilities, library support, bookstore, tutorial services, mediation and conflict resolution, social support network, student's newsletter, internship and employment services, alumni affairs, and other services. For distance learners, institutions should consider providing toll-free numbers for many of these services.

Pre-Enrollment Services

Institutions should provide information sessions through which learners can receive information about the course/program and all support services before registration.

Orientation

Remote learners always need detailed information about the course and usually appreciate any tips that can help them successfully complete the course. Therefore, there should not be any doubt that an e-learning course needs a more detailed orientation than a traditional classroom based course. E-learning students should be clearly informed about course expectations and requirements.

All students should be required to participate in an online orientation at least a week prior to the first day of classes. Orientation should provide introduction to procedures for learning at a distance, including roles and responsibilities of instructors, learners (Gibson, 1998), tutors, facilitators, guest speakers, and all other individuals involved in the process. Students, instructors, and technical staff should be encouraged to post brief biographies and this helps to create a virtual learning community (Khan, 1997). Spitzer (2001) states that one of the best ways to get students involved early on in a course is to ask them to share their expectations. At the orientation, students should be given appropriate login password to enter the course and also should be informed how to get identity cards (if required for the course). University of Maryland University College provides online orientation for distance students where they receive the overview of online learning and are able to get started (http://umuc.edu/distance/de_orien) with their e-learning. Northern Virginia Community College provides a Technology Guide for students that includes how to access online resources and online learning activities, information about hardware, and software related to online courses (http://www.nv.cc.va.us/resources/stutechguide/stutechguide.pdf).

Faculty and Staff Directories

Distance learners and instructors benefit from having faculty and staff directories online with e-mail addresses, telephone numbers, fax numbers, mailing addresses, Web pages, and so on.

Advising

Distance learners should receive academic and enrollment advising from qualified advisors on issues such as course selection, procedures for transfer credits, financial aids, institutional policies, available e-learning resources for students, and so on. Peer advising can be used effectively in e-learning. Students who completed some e-learning courses can be asked to volunteer as peer-guides for new students. At the Jones International University (JIU), a list of peer advising counselors is provided for new students. These peer guides who are known as "e-mail buddies" at JIU help new students in courses-related issues, requirements, and study tips.

Counseling

Like on-campus students, distance learners should receive career and other counseling services. Qualified counselors should be available to assist students in planning their academic programs, selecting appropriate courses for their program, and providing guidance on study skills, time management, stress management, and personal problems. At Cerro Coso Community College, students can contact a counselor directly via e-mail (http://www.cerrocoso.edu/studentservices/counseling/index.htm). More discussion about counseling can be found in the *Online Support* section of Chapter 8).

Learning Skills Development

A learner who has no experience in open, flexible, and distributed learning environment will greatly benefit from a learner's guide. In a traditional campus, students can go to a learning skill center to get help. In an e-learning environment, a well-designed learner's guide can serve as an effective learning center. The Open University of England provides learner's guide to help students develop skills in writing and examination skills (http://www3.open.ac.uk/learners-guide/learning-skills/index.htm).

Services for Students with Disabilities

Institutions offering e-learning should make sure that students with disabilities are able to access its courses and programs, and also receive adequate services for learning. Therefore, institutions should consider developing special services to help these students to succeed in their courses (issues concerning students with disabilities are also discussed in *Page and Site Design* section of Chapter 7, and *Learner Diversity* and *Information Accessibility* sections of Chapter 6). For example, at the Athabasca University, services are being developed to respond to a wide variety of needs of students with disabilities:

Students can presently receive information; assessments for assistive technology; assistance, and/or referral for funding and services; help with study skill and organizational strategies; extension of course contract dates; alternative methods for writing exams; and a variety of other services (http://www.athabascau.ca/html/services/advise/disab.htm).

The Open University of England provides services for students with following disabilities: Blind or impaired sight, deaf or hard of hearing, restricted mobility, restricted manual skills, dyslexic, or other specific learning difficulties, mental health difficulties, medical conditions, or impaired speech (http://www3.open.ac.uk/learners-guide/disability/sec_b/sec_b.htm).

Library Support

Both online and offline library support services are highly desirable by distance learners. Library support is one of the most critical support services for the success of e-learning. According to ACRL or Association of College & Research Libraries (2004), "Library resources and services in institutions of higher education must meet the needs of all their faculty, students, and academic support personnel, regardless of where they are located." ACRL developed the "Guidelines for Distance Learning Library Services" which emphasize that the library services offered to the distance learning community should be designed to meet effectively a wide range of informational, bibliographic, and user needs. The exact combination of central and site staffing for distance learning library services will differ from institution to institution. Therefore, institutions should plan for reliable and efficient library support services for e-learning (discussed in detail in the *Resources* section of Chapter 8). The following Web site has detailed and updated information about ACRL guidelines: http://www.ala.org/ala/acrl/acrlstandards/guidelinesdistancelearning.htm

Bookstore

An e-learning environment should be designed as a virtual campus where all possible students services, including online bookstore and online student union are included. If an institution has a campus bookstore, then the bookstore should consider having an online catalog where students can order books online. Students can also be directed to other online bookstores for purchasing their books. Cardean University has partnered with Specialty Books to create the Cardean University Bookstore where students can purchase textbooks, course packs, and software with a discount (http://www.specialty-books.com/cardean/).

Tutorial Services

Instructor (or academic advisor) should monitor students' performance during the course. Students who experience academic difficulties (i.e., struggling students) should receive academic assistance such as referring them to appropriate persons for assistance or providing tutorials. University of Phoenix provides multi-media tutorials and simulations through a customized Web portal (https://ecampus.phoenix.edu/tutorials.asp).

Mediation and Conflict Resolution

Institutions offering e-learning should make sure that all distance learners receive fair and equitable treatment from their faculty and administrative staff. Like traditional campus-based educational systems, contentious issues in e-learning environments can include academic grievances, grade disputes, student/instructor conflicts, sexual harassment, and discrimination concerns. Distance learners should be able to contact individuals (commonly known as ombuds officers) who can investigate their complaints and help them achieve fair and reasonable settlements. For example, Athabasca University has Ombuds staffers who assist students in finding solutions to problems concerning various students' services (http://www.athabascau.ca/studserv/ombuds.html).

Social Support Network

Institution can develop social support networks to help distance learners and instructors overcome the isolation that often accompanies distance learning (Dirr, 1999). Online student unions and teachers' lounges can be created to support the social network. Faculty and students can share their thoughts, problems and solutions regarding online education in their respective lounges. Walden University hosts student union (http://www.waldenu.edu/union/) and faculty lounge (http://www.waldenu.edu/faculty/) Web sites.

Student Newsletter

Institution can publish daily, weekly, monthly, quarterly, or semesterly online newsletters with information important to learners. Open Learning Australia provides "Dialogue", a student newsletter that contains items of interest to students, student profiles, news on new units and modules, and any changes that have been made to textbooks. There are also opportunities for students to contribute and occasional competitions (http://www.ola.edu.au/resources/support.asp#dialogue). An online education newsletter that provides the directory of online courses, online learning and online education is located at (http://www.onlineeducationnewsletter.com/) might be of interest to online learners.

Internship and Employment Services

Institutions should create or link to online information sites for internship and job opportunities for online students and alumni. Northern Virginia Community College directs students and alumni from its Web site to http://www.jobtrak.com (a commercial site with job listings and resume databases).

Alumni Affairs

Alumni (or past students) play a very critical role for their alma mater. Alumni can not only help their alma mater recruit new students, but also can help students in their learning activities by volunteering as mentors and by pointing students to both internship and job opportunities. Institutions should host a database on the Internet with listing of alumni and current students. This database will help all past and present students to communicate with each other. Current students can seek advice from past students about career planning. Institutions can host discussion forums to foster communications among alumni and currents students. Capella University hosts an Alumni Discussion Forum moderated by the Capella Alumni Association that provides an opportunity for its graduates to post their professional comments, questions, and suggestions (http://courses.capellauniversity.edu/capella/discussion/alumdisc.nsf). Walden University Alumni Association hosts a Web site at http://www.waldenu.edu/alumni/.

Other Services

As indicated in Chapter 1, e-learning is new to many students. E-learning providers should continue to finds innovative ways to provide various services for distance learners. They should conduct ongoing surveys to learn about the effectiveness of exiting services and inquire about new services that might be needed.

Question to Consider

Can you think of any e-learning related academic, administrative, and student services issues not covered in this chapter?

Activity

1.   Using Internet search engines, locate at least one article covering any of the following student services issues, and analyze the article from the perspective of its usefulness in e-learning:

- Pre-enrollment services
- Orientation
- Faculty and staff directories
- Advising
- Counseling
- Learning skills development
- Services for students with disabilities

- Library support
- Bookstore
- Tutorial services
- Mediation and conflict resolution
- Social support network
- Student newsletter
- Internship and employment services
- Alumni affairs
- Other services

2. Locate an online program and review its *institutional affairs* using the relevant *institutional checklist items* at the end of Chapter 2.

References

Accetta, R. (2001). Courses for profit: Does e-learning abuse academic labor? *e-learning, 2*(8), 26-30.

Aggarwal, A. (ed.) (2000). *Web-based learning and teaching technologies: Opportunities and challenges*. Hershey, PA: Idea Group Publishing.

Bates, T. (2001). International distance education: Cultural and ethical issues. *Distance Education: An International Journal, 22*(1), 122-136.

Belanger, F. & Jordan, D.H. (2000). *Evaluation and implementation of distance learning: Technologies, tools, and techniques*. Hershey, PA: Idea Group Publishing.

Berge, Z.L. (2000). Evaluating Web-based training programs. In B.H. Khan (Ed.), *Web-based training* (pp. 515-522). Englewood Cliffs, NJ: Educational Technology Publications.

Bischoff, A. (2000). The elements of effective online teaching. In K.W. White & B.H. Weight (Eds.), *The online teaching guide*. Needham Height, MA: Allyn & Bacon.

Boettcher, J.V. (1999). How much does it cost to develop a distance-learning course? It all depends… [online]. *http://www.cren.net/~jboettch/dlmay.htm*

Chapnick, S. (2001). Are you ready for e-learning? Retrieved January 24, 2003, from *http://www.learningcircuits.org/nov2000/chapnick.html*

Connick, G.P. (1998). Choosing a distance education provider: Asking the right questions. In G.P. Connick (Ed.). *The distance learner's guide*. Upper Saddler River, NJ: Prentice Hall.

Dirr, P.J. (1999). *Putting principles into practice: Promoting effective support services for students in distance learning programs*. A report on the findings of a survey. Public Service Telecommunications Corporation.

Ely, D.P. (1999). Conditions that facilitate the implementation of educational technology innovations. *Educational Technology*, November-December, 23-27.

Gibson, C.C. (1998). In Retrospect. In C.C. Gibson (Ed.), *Distance learners in higher education*. Madison, Wisconsin: Atwood Publishing.

Gustafson, K.L. & Schrum, L. (2000). Cost analysis and return on investment (ROI) for distance education. In B.H. Khan (Ed.), *Web-based training* (pp. 537-558). Englewood Cliffs, NJ: Educational Technology Publications.

Jennings, M.M. & Dirksen, D.J. (1997). Facilitating change: A process for adoption of Web-based instruction. In B.H. Khan (Ed.), *Web-based instruction* (pp. 111-118). Englewood Cliffs, NJ: Educational Technology.

Kearsley, G. (2002). Is online learning for everybody? Retrieved June, 2004, from *http://home.sprynet.com/~gkearsley/everybody.htm*

Kearsley, G. & Marquardt, M.J. (2001). Infostructures: Technology, learning, and organizations. In B.H. Khan (Ed.), *Web-based training,* (pp. 27-32). Englewood Cliffs, NJ: Educational Technology Publications.

Khan, B.H. (1997). Web-based instruction: What is it and why is it? In B.H. Khan (Ed.), *Web-based instruction,* (pp. 5-18). Englewood Cliffs, NJ: Educational Technology Publications.

Levenberg, N. (2001). Positioning for effectiveness: Applying marketing concepts to Web-based training. In B. H. Khan (Ed.), *Web-based training,* (pp. 395-398). Englewood Cliffs, NJ: Educational Technology Publications.

Morgan, C.K. & Tam, M. (1999). Unraveling the complexities of distance education student attrition. *Distance education, 20*(1), 96-108.

Passmore, D.L. (2000). Impediments to adoption of Web-based course delivery among university faculty. The Pennsylvania State University. Retrieved January 24, 2003, from *http://www.aln.org/alnweb/magazine/ Vol4_issue2/passmore.htm*

Price, R.V. (1999). Designing a college Web-based course using a modified personalized system of instruction (PSI) model. *TechTrends, 43*(5), 23-28.

Romiszowski, A.J. & Chang, E. (2001). A practical model for conversational Web-based training: A response from the past to the needs of the future. In B.H. Khan (Ed.), *Web-based training,* (pp. 107-128). Englewood Cliffs, NJ: Educational Technology Publications.

Rossner, V. & Stockley, D. (1997). Institutional perspectives on organizing and delivering Web-based instruction. In B.H. Khan (Ed.), *Web-based instruction,* (pp. 333-336). Englewood Cliffs, NJ: Educational Technology.

Schelin, E. (2001). Profiles in e-learning. An e-learning visionary pushes the envelope. *e-learning, 2*(6), 48-50.

Spitzer, D.R. (2001). Don't forget the high-touch with the high tech in distance learning. *Educational Technology*, *41*(2), 51-55.

Stacey, P. (2001). Learning Management Systems (LMS) & Learning Content Management Systems (LCMS) - E-learning an enterprise application? Retrieved January 24, 2003, from *http://www.bctechnology.com/statics/pstacey-oct2601.html*

Twigg, C.A. (2000). Who owns online courses and course materials? Intellectual property policies for a new learning environment. Retrieved January 24, 2003, from *http://www.center.rpi.edu/PewSym/mono2.html*

Welsch, E. (2002). Cautious steps ahead. *Online Learning Magazine, 6*(1), 20-24.

Williams, V. & Peters, K. (1997). Faculty incentives for the preparation of Web-based Instruction. In B.H. Khan (Ed.), *Web-based instruction,* (pp. 107-110). Englewood Cliffs, NJ: Educational Technology Publications.

Institutional Checklist

Needs Assessment

Does the institution conduct a survey to identify whether e-learning is suitable for learners?

❑   Yes

❑   No

❑   Not applicable

❑   Other (specify)

Does the institution conduct a need analysis for e-learning?

❑   Yes

❑   No

❑   Not applicable

❑   Other (specify)

If *yes*, check all that apply:

❑   To identify e-learning needs

- ❏ To identify technological needs for e-learning environment
- ❏ To identify other support services needs for e-learning environment
- ❏ To identify customers' (i.e., learners') willingness to take e-learning courses from the institution
- ❏ Not applicable
- ❏ Other (specify)

Financial Readiness

Is the e-learning initiative aligned with the institution's mission?
- ❏ Yes
- ❏ No
- ❏ Not applicable
- ❏ Other (specify)

Is the institution financially ready to venture into e-learning?
- ❏ Yes
- ❏ No
- ❏ Not applicable
- ❏ Other (specify)

Does the e-learning initiative have direct support from the senior administrative staff of the institution?
- ❏ Yes
- ❏ No
- ❏ Not applicable
- ❏ Other (specify)

Is the e-learning initiative dependent on financial sources?
- ❏ Yes
- ❏ No
- ❏ Not applicable
- ❏ Other (specify)

If *yes*, check all the apply:

- ☐   Internal funds from the institution
- ☐   Tuition and fees from students
- ☐   External funds from (check all that apply)
  - ☐   Federal or national government
  - ☐   State or provincial government
  - ☐   County or district government
  - ☐   Industry
  - ☐   Foundation
  - ☐   Other (specify)

Does the institution have adequate funds for e-learning?

- ☐   Yes
- ☐   No
- ☐   Not applicable
- ☐   Other (specify)

Does the institution have a timeline associated with how funds will be available during various phases of the e-learning initiative?

- ☐   Yes
- ☐   No
- ☐   Not applicable
- ☐   Other (specify)

If funds are not available as expected, does the institution have a plan that can eliminate not-so-critical areas of the e-learning initiative?

- ☐   Yes
- ☐   No
- ☐   Not applicable
- ☐   Other (specify)

Infrastructure Readiness

Does the institution have adequate human resources to support the e-learning initiative?

❑ Yes
❑ No
❑ Not applicable
❑ Other (specify)

If yes, check all that apply:
❑   Help desk staff
❑   Technical support staff
❑   Training staff to train e-learning staff
❑   Other (specify)

Does the institution have adequate equipment to support the e-learning initiative?
❑ Yes
❑ No
❑ Not applicable
❑ Other (specify)

Check appropriate hardware available for the role of each individual listed below. Check all that apply (put "NA" if not applicable).

Role of Individual	Hardware																					
	Desktop Computer			Laptop Computer			PDA			Pager / Cellular Telephone			Printer / Scanner			Digital Camera / WebCam			Other (specify)		
	Work	Home	Mobil	Work	Home	Mobil	Work	Home	Mobil	Work	Home	Mobil	Work	Home	Mobil	Work	Home	Mobil	Work	Home	Mobil
Learner																						
Instructor (full-time)																						
Instructor (part-time)																						
Trainer																						
Trainer Assistant																						
Tutor																						
Technical Support																						
Help Desk																						
Librarian																						
Counselor																						
Graduate Assistant																						
Administrator																						
Other (specify)																						

Check appropriate Internet connection types available for the role of each individual listed below. Check all that apply (put "NA" if not applicable).

Role of Individual	Internet Connections																					
	Dial-Up			ISDN			Satellite Dish			Cable Modem			DSL			T1			Wireless / Other (specify)		
	Work	Home	Mobile	Work	Home	Mobile	Work	Home	Mobile	Work	Home	Mobile	Work	Home	Mobile	Work	Home	Mobile	Work	Home	Mobile
Learner																					
Instructor (full-time)																					
Instructor (part-time)																					
Trainer																					
Trainer Assistant																					
Tutor																					
Technical Support																					
Help Desk																					
Librarian																					
Counselor																					
Graduate Assistant																					
Administrator																					
Other (specify)																					

Check appropriate computer skills and familiarity with computer technology by the role of each individual listed below. Check all that apply (put "NA" if not applicable).

Role of Individual	Computer Skills and Computer Usage																				
	Word Processing			Internet Connectivity			Browsing			Familiarity with the Use of Audio and Video on the Internet			Familiarity with Computer Terms and Jargon			Years of Computer Usage			Other (specify)		
	Yes	No	NA	Yes	No	NA	Yes	No	NA	Yes	No	NA	None	1-2	2-5	Yes	No	NA	Yes	No	NA
Learner																					
Instructor (full-time)																					
Instructor (part-time)																					
Trainer																					
Trainer Assistant																					
Tutor																					
Technical Support																					
Help Desk																					
Librarian																					
Counselor																					
Graduate Assistant																					
Administrator																					
Other (specify)																					

Check the availability of appropriate hardware and software by the institution for e-learning. Check all that apply:

Hardware and Software	Availability			
	Yes	No	NA	Other (specify)
Servers				
Database				
Learning Management System (LMS)				
Learning Content Management System (LCMS)				
Enterprise Software				
Other (specify)				

How much experience does any of the following have with the online learning environment? Check all that apply:

Role of Individual	Experience Level			
	None	Some	Adequate	Other (specify)
Learner				
Instructor (full-time)				
Instructor (part-time)				
Teaching Assistant				
Tutor				
Technical Support				
Help Desk				
Librarian				
Counselor				
Graduate Assistant				
Administrator				
Other (specify)				

Does the institution use any of the following methods to identify the assistive technology and other support that disabled students may need to participate in e-learning? Check all that apply:

Methods	Yes	No	NA	Other (specify)
Interview				
Observation				
Document Review				
Survey				
Other (specify)				

Does the institution have a plan to train its staff for any new technological skills that they might need in the future?

❑   Yes

❑   No

❑   Not applicable

❑   Other (specify)

If *yes*, check all that apply:

❑   Full-time training staff

❑   Part-time training staff

❑   Planned full-time training staff

❑   Planned part-time training staff

❑   Staff are sent to get training from outside

❑   Outside consultants train its staff

❑   Other (specify)

If necessary technological skills are lacking, what steps will the institution take to speed up acquiring the skill?

Cultural Readiness

Is any of the following ready for e-learning? (Note: It may be beneficial for the institution to name individual(s) or appoint a group responsible for identifying the cultural readiness of the following individuals.)

Role of Individual	Ready for E-Learning			
	Yes	No	NA	Other (specify)
Learner				
Instructor (full-time)				
Instructor (part-time)				
Teaching Assistant				
Tutor				
Technical Support				
Help Desk				
Librarian				
Counselor				
Graduate Assistant				
Administrator				
Other (specify)				

Check all that apply for individuals' preferences for learning format.

Role of Individual	Preferences											
	On-line Learning (OL)			*Face-to-Face Instruction (F2F)*			*Combination of OL, F2F and Other (Blended)*			*Other (specify)*		
	Yes	No	NA	Yes	No	NA	Yes	No	NA	Yes	No	NA
Learner												
Instructor (full-time)												
Instructor (part-time)												
Teaching Assistant												
Tutor												
Technical Support												
Help Desk												
Librarian												
Counselor												
Graduate Assistant												
Administrator												
Other (specify)												

Content Readiness

Is content ready? (Note: Inventory should specify the location and responsible authors/producers for all multimedia contents, and indicate if permission is needed from copyright holders.)

❑   Yes

❑   No

❑   Not applicable

❑   Other (specify)

If *yes*, what percent of content is ready?

Percent (%) Completed	Content Types					
	Text	*Graphics*	*Video*	*Audio*	*Animation*	*Other (specify)*
None						
Less than 50%						
More than 50%						
Other (specify)						

Organization and Change

Check if the institution developed any of the following for its e-learning initiative.
- ❏ Mission Statement
- ❏ Strategic Plan
- ❏ Business Plan
- ❏ Not applicable
- ❏ Other (specify)

To what extent there is buy-in from key players within the institution? Please describe for each key player.

Has the institution made any change in its organizational structure to accommodate the needs of open, flexible, and distributed eleaning?
- ❏ Yes
- ❏ No
- ❏ Not applicable
- ❏ Other (specify)

   If *yes*, please explain below:

Diffusion and Adoption

Does the institution inform the stakeholders (i.e., learners, instructors, support services staff and community members) about the benefits of an e-learning initiative?
- ❏ Yes
- ❏ No
- ❏ Not applicable
- ❏ Other (specify)

Does the institution have a system of keeping all stakeholder groups informed about the activities of its e-learning initiative?

❑   Yes

❑   No

❑   Not applicable

❑   Other (specify)

If *yes*, check all that apply:

Stakeholders	Method of Communication (Rate each method from 1-10 scale where 10 represents the most informed and 1 represents the least informed.)				
	E-Mail	*Listserv or Discuss Forum*	*Newsletter*	*Community Newspaper*	*Other (specify)*
Learners					
Instructors					
Support Staff					
Management					
Other (specify)					

Does the institution provide any of the following incentives for any of the following individuals involved in e-learning initiatives?

❑   Yes

❑   No

❑   Not applicable

❑   Other (specify)

If *yes*, check all that apply:

Role of Individual	Incentives			
	Release Time	*Financial Reward*	*Tenure*	*Other (specify)*
Instructor				
Trainer				
Teaching Assistant				
Training Assistant				
Tutor				
Project Manager				
Counselor				
Librarian				
Other (specify)				

Implementation

Does the institution clearly identify skills and knowledge required by learners to adopt e-learning?

☐   Yes

☐   No

☐   Not applicable

☐   Other (specify)

Does the institution provide reward or incentives for learners to adopt e-learning?

☐   Yes

☐   No

☐   Not applicable

☐   Other (specify)

If *yes*, please explain below:

Will the key leaders of the institution be involved in day-to-day activities of the e-learning initiative?

☐   Yes

☐   No

☐   Not applicable

☐   Other (specify)

If *yes*, please explain below:

Does the institution's e-learning initiative receive continuing support from key individuals in the institution?

☐   Yes

□   No
□   Not applicable
□   Other (specify)

Budgeting and Return on Investment

Does the institution have a budget for e-learning?
□   Yes
□   No
□   Not applicable
□   Other (specify)

Check if the institution has budgeted for any of the following costs related to e-learning. Check all that apply:
□   Content expert
□   Course designers (i.e., instructional designers)
□   Computer programmers
□   Graphic artists
□   Instructor compensation
□   Support staff compensation
□   Consultants
□   Equipment and software acquisition
□   Marketing
□   Delivery
□   Program revision
□   Ongoing maintenance and upgrade of equipment and software
□   Operating costs such as telephone, postage and supplies
□   Not applicable
□   Other (specify)

Formula for Return on Investment (ROI)

$$\text{ROI (\%)} = \frac{\text{Benefits} - \text{Total Costs}}{\text{Total Costs}} \times 100$$

$$ROI\,(\%)\;=\;\frac{\text{Net Benefits}}{\text{Costs}}\;\;X\;\;100$$

Does the institution conduct return on investment (ROI) analysis during any of the following stages of e-learning?

❑   Yes

❑   No

❑   Not applicable

❑   Other (specify)

    If *yes*, check all that apply:

    ❑   Before implementation

    ❑   During implementation

    ❑   After implementation

    ❑   Other (specify)

Does the institution put aside any additional money for obtaining any of the following resources that may be needed in the future to continue e-learning?

❑   Yes

❑   No

❑   Not applicable

❑   Other (specify)

    If *yes*, check all that apply:

    ❑   Hardware

    ❑   Software

    ❑   Not applicable

    ❑   Other (specify)

Partnership with Other Institutions

Does the institution have partnerships with other e-learning institutions?

❑   Yes

❑   No

❑   Not applicable
❑   Other (specify)

    If *yes*, check all that apply:
    ❑   Students can take courses from partner institutions toward a degree
        program
    ❑   Students can transfer courses within partner institutions
    ❑   Not applicable
    ❑   Other (specify)

Can students use library and other learning resources from partner institutions?
❑   Yes
❑   No
❑   Not applicable
❑   Other (specify)

Program and Course Information Catalogue

Check if the institution provides its program and course information via any of
the following formats:
❑   Completely online
❑   Completely print materials
❑   Partially online
❑   Other (please describe)

What is the format of the course?
❑   All online
❑   Partially online

    If partially online, what other media are used?
    ❑   Face-to-face classes
    ❑   CD ROMs
    ❑   DVDs
    ❑   Interactive TV

❑   Satellite

❑   Printed materials

❑   Other (please describe)

Are students required to take a prerequisite course before taking an "online" course from the institution?

❑   Yes

❑   No

❑   Not applicable

❑   Does not specify

❑   Other (specify)

Is the course a part of a specific degree (or certification) program?

❑   Yes

❑   No

❑   Not applicable

❑   Other (specify)

If *yes*, check all that apply:

❑   Required course

❑   Elective course

❑   Other (please describe)

Does the institution provide information about whether the course can be used toward degree programs in various fields?

❑   Yes

❑   No

❑   Not applicable

❑   Other (specify)

If *yes*, list the name(s) of the academic or professional programs.

Does the course allow students to preview any part of course materials (or course demo) before registration?

❑   Yes

❑   No

❑   Not applicable

❑   Other (specify)

Does the institution provide any information regarding whether the course is transferable to other accredited institutions?

❑   Yes

❑   No

❑   Not applicable

❑   Other (specify)

Does the institution post updated information about the courses that are currently unavailable for registration?

❑   Yes

❑   No

❑   Not applicable

❑   Other (specify)

Does the institution provide information about the accreditation status of the course and the institution?

❑   Yes

❑   No

❑   Not applicable

❑   Other (specify)

Are students required to be part of a degree program to take this course?

❑   Yes

❑   No

❑   Not applicable

❑   Other (specify)

Academic Calendar and Course Schedule

What is the course schedule format?

- ❑   Fixed start and fixed end date
- ❑   Fixed duration (e.g., must be finished within a semester or a year, etc.)
- ❑   Open (can start any time and finish any time)
- ❑   Not applicable
- ❑   Other (specify)

Provide information about the course schedule format:

Quarter / Semester	Year	Course Name	Registration Dates	Course Start Date	Course End Date
Fall					
Winter					
Spring					
Summer					
Other (specify)					

Provide information about the synchronous (or live) activities scheduled for the course:

Live Activities	Date (e.g., May 23, 2004)	Start Time* (e.g., 16:00)	End Time* (e.g., 17:00)
Live Chat			
Audio Conferencing			
Video Conferencing			
Telephone			
Other (specify)			

*UTC (Universal Time Coordinated) which is equivalent to GMT (Greenwich Mean Time) should also be provided for learners located in different time zones.

Check the course duration.

- ❑   Intensive (1-2 months duration)
- ❑   Quarter-based (3 months duration)
- ❑   Semester-based (4 months duration)
- ❑   Independent-study (12 months duration)

❑   Self-paced
❑   Other (specify)

What is length (i.e., instructional hours) of the course? (Note: The length of the course can be dependent on learners' stated needs or interests. Learners can work on their own paces. For example, they can begin in January and ends in June.)

Tuition and Fees

Does the institution provide information about course tuition and fees? If *yes*, enter the amount of money in the appropriate boxes below:

Tuition and Fees	Yes	No	NA	US $*
Course Tuition (per credit)				
Technology Fee (per course)				
Registration Processing Fee				
Application Fee for Admission				
Graduation Fee				
Official Transcript				
Graduation Certificate				
Charge for Returned Check				
Other Charges (specify)				

* [If appropriate, use your own currency].

How does *tuition* for online students compare to tuition for facetoface students in the same institution?
❑   Same
❑   Higher
❑   Lower
❑   Not applicable
❑   Other (specify)

Are online courses offered at a lower *fee* than on-campus courses? (Note: It is not uncommon for students to expect lower fees for online courses since they do not use a physical classroom.)
❑   Yes
❑   No
❑   Not applicable
❑   Other (specify)

How does *tuition* compare to that of *similar institutions*?
- ❑ Same
- ❑ Higher
- ❑ Lower
- ❑ Not applicable
- ❑ Other (specify)

How do *fees* compare to those of *similar institutions*?
- ❑ Same
- ❑ Higher
- ❑ Lower
- ❑ Not applicable
- ❑ Other (specify)

How does *tuition* compare to that of *partner institutions*?
- ❑ Same
- ❑ Higher
- ❑ Lower
- ❑ Not applicable
- ❑ Other (specify)

How do *fees* compare to that of *partner institutions*?
- ❑ Same
- ❑ Higher
- ❑ Lower
- ❑ Not applicable
- ❑ Other (specify)

Does the institution provide a Withdrawal Request Form (WRF) online?
- ❑ Yes
- ❑ No
- ❑ Not applicable
- ❑ Other (specify)

If *yes*, check all that apply:

- ❑ Students can submit WRF Online
- ❑ Students can fax WRF
- ❑ Students can send WRF via regular mail
- ❑ Other (specify)

Does the course provide clear information about refund policies?

- ❑ Yes
- ❑ No
- ❑ Not applicable
- ❑ Other (specify)

If *yes*, enter % of reimbursement of tuition in the appropriate places below:

Withdrawal Date	Percentage of Tuition Refunded
1st Week	
2nd Week	
3rd Week	
4th Week	
5th Week	
6th Week	
Other (specify)	

May the institution change fees without notice?

- ❑ Yes
- ❑ No
- ❑ Not applicable
- ❑ Other (specify)

Marketing and Recruitment

Does the institution offer information sessions to recruit students?

- ❑ Yes
- ❑ No
- ❑ Not applicable
- ❑ Other (specify)

If *yes*, check all the apply:
- ❑   Online information session
- ❑   Face-to-face information sessions
- ❑   Not applicable
- ❑   Other (specify)

Check if the institution has any of the following for its e-learning marketing team. Check all the apply:
- ❑   Market researcher
- ❑   Recruiter
- ❑   Not applicable
- ❑   Other (specify)

Does the course offer features (e.g., attractive design, quick loading time, comprehensive content, multi-cultural, etc.) that make it highly marketable?
- ❑   Yes
- ❑   No
- ❑   Not applicable
- ❑   Other (specify)

Does the institution market its e-learning offerings?
- ❑   Yes
- ❑   No
- ❑   Not applicable
- ❑   Other (specify)

If *yes*, indicate if the institution uses any of the following means to market its e-learning offerings. Check all the apply:
- ❑   Websites
- ❑   Internet banner ads
- ❑   E-mail
- ❑   Listserv postings
- ❑   Newsletter
- ❑   Newspaper

❑   Magazine

❑   Journal

❑   Radio

❑   TV

❑   Posters

❑   Free promotional gifts

❑   No applicable

❑   Other (specify)

Does the institution provide previous students' testimonials about its e-learning courses and programs?

❑   Yes

❑   No

❑   Not applicable

❑   Other (specify)

Does the institution provide information about student attrition (or dropout) rate? (Note: Institutions need a process and a person responsible for analyzing attrition and creating strategies to reduce it.)

❑   Yes

❑   No

❑   Not applicable

❑   Other (specify)

Does the institution provide any statistics on how many students successfully completed the online course?

❑   Yes

❑   No

❑   Not applicable

❑   Other (specify)

Does the institution provide any statistics on the types of barriers causing attrition?

❑   Yes

❑   No

❑   Not applicable

❑   Other (specify)

If *yes*, enter % dropped due to each barrier in the appropriate boxes below:

Barriers	Percent (%) Dropped for the Barrier
Lack of free time	
Change in circumstances	
Took more time than expected	
Personal study problems	
Unclear goals	
Time management problems	
Problem with course schedule and pacing	
Learning materials arrive late	
Insufficient feedback on assignments	
Insufficient/unsatisfactory communication with instructional staff	
Course focus and expectations not clear	
Inflexible course structure	
Course content outdated	
Confusing changes to course	
Unresponsive instructor	
Inadequate technical support	
Inadequate library support	
Difficult content	
Mismatch in assessment requirement	
Course focus lacked personal relevance or interest	
Lacked prerequisite knowledge	
Other (specify)	

Admissions

Does the institution provide admission requirements information online?

❑   Yes

❑   No

❑   Not applicable

❑   Other (specify)

Do students have access to individual(s) at the admission office?

❑   Yes

❑   No

❑   Not applicable
❑   Other (specify)

If *yes*, indicate the means of communication (check all that apply):
❑   Phone
❑   E-mail
❑   Online (realtime)
❑   Other (specify)

Does the institution provide information about transfer credit requirements?
❑   Yes
❑   No
❑   Not applicable
❑   Other (specify)

Check if students can submit Admission Application Form via any of the following methods. Check all that apply:
❑   Online
❑   Regular mail
❑   Other (specify)

Can students submit all required admission forms online?
❑   Yes
❑   No
❑   Not applicable
❑   Other (specify)

Can students inquire about the status of their application online?
❑   Yes
❑   No
❑   Not applicable
❑   Other (specify)

Financial Aid

Can students apply for financial aid online?
- ❑   Yes
- ❑   No
- ❑   Not applicable
- ❑   Other (specify)

Are students informed about how soon they receive financial aid eligibility?
- ❑   Yes
- ❑   No
- ❑   Not applicable
- ❑   Other (specify)

Do online students receive the same financial aid programs as on-campus students?
- ❑   Yes
- ❑   No, they do not receive any financial aid
- ❑   No, they have special financial aid programs
- ❑   Not applicable
- ❑   Other (specify)

Does the institution offer federal work-study or community service opportunities for online students?
- ❑   Yes
- ❑   No
- ❑   Not applicable
- ❑   Other (specify)

Does the institution have any scholarships available for online students?
- ❑   Yes
- ❑   No
- ❑   Not applicable
- ❑   Other (specify)

Does the institution provide links to scholarship Websites? (Note: In the USA, sites such as http://www.wiredscholar.com provide scholarship information based on location, age, school year, and heritage of applicants.)

❏   Yes

❏   No

❏   Not applicable

❏   Other (specify)

Registration and Payment

Can students register online?

❏   Yes

❏   No

❏   Not applicable

❏   Other (specify)

If *no*, check other type(s) of registration methods available for the course:

❏   By regular mail

❏   By telephone call to a person doing registration

❏   By fax

❏   By touch-tone phone

❏   Other (specify)

Does the institution have an online system that gives students a secure method of making their tuition and fees payment?

❏   Yes

❏   No

❏   Not applicable

How do students pay tuition and fees? (Check all that apply):

Methods of Payment (Tuition and Fees)	Transaction	
	On-line	Off-line
Credit cards		
Checks		
Electronic money transfer		
Money order		
Electronic checks (e.g., Western Union in the USA)		
Automatic debit system		
Gift certificates or coupons		
Other (specify)		

If registration is done online, does the student receive confirmation of registration?

❑   Yes

❑   No

❑   Not applicable

If *yes*, check all that apply:

❑   Confirmation receives online (e.g., e-mail, online message, etc.)

❑   Confirmation receives off-line (e.g., letter, phone, etc.)

❑   Other (specify)

Does the institution have a system for providing online billing information?

❑   Yes

❑   No

❑   Not applicable

❑   Other (specify)

Does the institution provide online procedures for dropping or withdrawing from an e-learning course?

❑   Yes

❑   No

❑   Not applicable

❑   Other (specify)

If *yes*, is any penalty involved?

- ❏ Yes
- ❏ No
- ❏ Not applicable
- ❏ Other (specify)

If the course is offered in modules at different times, does the institution allow students the flexibility to enroll for their desired module(s)?

- ❏ Yes
- ❏ No
- ❏ Not applicable
- ❏ Other (specify)

Would student information submitted online to the registrar's office be kept secure and confidential to the extent possible?

- ❏ Yes
- ❏ No
- ❏ Not applicable

Does the institution provide enrollment history of the course (i.e., how many students enrolled in the course in past semesters)?

- ❏ Yes
- ❏ No
- ❏ Not applicable
- ❏ Other (specify)

Does the institution provide the completion history of the course (i.e., what percent of enrolled students completed the course)?

- ❏ Yes
- ❏ No
- ❏ Not applicable
- ❏ Other (specify)

Does the course offer a money-back guarantee?

- ❑   Yes
- ❑   No
- ❑   Not applicable
- ❑   Other (specify)

Is the following registration-related information for the course provided?

- ❑   Total number of seats allocated for the course
- ❑   Number of seats available
- ❑   Not applicable
- ❑   Other (specify)

Information Technology Services

Do students have any of the following facilities or support from the information technology services?

- ❑   Yes
- ❑   No
- ❑   Not applicable
- ❑   Other (specify)

   If *yes*, check all the apply:
   - ❑   Borrow a laptop
   - ❑   Borrow a PDA (Personal Digital Assistant)
   - ❑   Borrow technical computer books/manuals
   - ❑   Borrow a scanner
   - ❑   Borrow a digital camera
   - ❑   Borrow a video camera
   - ❑   Borrow a digital video camera
   - ❑   Borrow an e-book reader
   - ❑   Borrow software
   - ❑   Receive spaces on the servers for personal Web pages
   - ❑   Receive e-mail account
   - ❑   Technical support

❑   Help desk
❑   Not applicable
❑   Other (specify)

Is the institution's information technology or computing services department involved in the selection of a learning management system (LMS) and/or learning content management system (LCMS) for e-learning?

❑   Yes
❑   No
❑   Not applicable
❑   Other (specify)

If *yes*, check if inputs from any of the following are considered in the LMS and LCMS selection process. Check all the apply:
❑   Instructional designers
❑   Content expert
❑   Instructors
❑   Students
❑   Technical support staff
❑   Library support staff
❑   Help desk
❑   Other (specify)

Instructional Design and Media Services

Does the institution have any of the following individuals involved in the creation of e-learning materials?
❑   Yes
❑   No
❑   Not applicable

If *yes*, check all the apply:
❑   Content or subject matter expert
❑   Instructional designer

❑   Programmer
❑   Multimedia designers
❑   Graphic artists
❑   Not applicable
❑   Other (specify)

Graduation

Does the institution provide an online means for tracking course completion in a program of study?

❑   Yes
❑   No
❑   Not applicable
❑   Other (specify)

Does the institution provide graduation information online?

❑   Yes
❑   No
❑   Not applicable
❑   Other (specify)

Does the institution update students about whether they completed all required courses toward a degree or not?

❑   Yes
❑   No
❑   Not applicable
❑   Other (specify)

Can students check the completion status of each course in their program plans before completing Graduation Application Form?

❑   Yes
❑   No
❑   Not applicable
❑   Other (specify)

If *yes*, check all that apply:
- ❑ Students can access their program plans online
- ❑ Students can call the admissions office to check the status of their program plan
- ❑ Admission or graduation office can e-mail their latest program plan to students
- ❑ Admission or graduation office can mail their latest program plan to students
- ❑ Other (specify)

Does the institution provide Graduation Application Form (GAF) online?
- ❑ Yes
- ❑ No
- ❑ Not applicable
- ❑ Other (specify)

If *yes*, check all that apply:
- ❑ Students can submit the GAF Online
- ❑ Students can fax the GAF
- ❑ Students can send the GAF via regular mail
- ❑ Other (specify)

Does the institution host a graduation ceremony for students who complete their programs totally online?
- ❑ Yes
- ❑ No
- ❑ Not applicable
- ❑ Other (specify)

If *yes*, check all that apply:
- ❑ Cyber Graduation Ceremony
- ❑ Onsite Graduation at the campus
- ❑ Other (specify)

Does the institution provide a Certificate Request Form (CRF) online?

❑   Yes

❑   No

❑   Not applicable

❑   Other (specify)

If *yes*, check all that apply:

❑   Students can submit a CRF online

❑   Students can fax a CRF

❑   Students can send a CRF via regular mail

❑   Other (specify)

Transcripts and Grades

Does the course provide the date for reporting the final grade?

❑   Yes

❑   No

❑   Not applicable

❑   Other (specify)

Does the course provide students' grades online?

❑   Yes

❑   No

❑   Not applicable

If *yes*, check all that apply:

❑   Weekly grading summary

❑   Mid-term grading summary

❑   Final grading summary

❑   Other (specify)

Does the institution provide a Transcript Request Form (TRF) online?

❑   Yes

❑   No
❑   Not applicable

If *yes*, check all that apply:
❑   Students can submit a TRF online
❑   Students can fax a TRF
❑   Students can send a TRF via regular mail
❑   Other (specify)

Does the institution provide information about the costs for official transcripts?
❑   Yes
❑   No
❑   Not applicable

If *yes*, enter the costs for official transcripts in the appropriate boxes:

Transcript Delivery	Cost per Transcript (within the country)	Cost per Transcript (outside the country)
Regular Mail		
Fax		
24-hour Mail		
48-hour Mail		
Other (specify)		

Accreditation

Is the institution accredited?
❑   Yes
❑   No
❑   Not applicable
❑   Other (specify)

If *yes*, indicate the name of the accrediting agency:

If *no*, check all that apply:
❑   Never applied for accreditation
❑   Application Pending

❑   Previously denied, applied again
❑   Other (specify)

Policies

Does the institution have e-learning policies and guidelines?
❑   Yes
❑   No
❑   Not applicable
❑   Other (specify)

Instructional Quality

Does the institution provide a list of mentors and/or tutors who are available to assist online students?
❑   Yes
❑   No
❑   Not applicable
❑   Other (specify)

Does the instructional staff maintain scheduled office hours for students?
❑   Yes
❑   No
❑   Not applicable
❑   Other (specify)

If yes, check the number of hours *per week*:

Instructional Staff	Hours /Week				
	5	10	15	20	Other
Instructor					
Tutor					
Other (specify)					

Check if students can contact instructional staff using any of the following methods during the posted office hours. Check all that apply:

Instructional Staff	Communication Methods					
	Phone	E-mail	Chat Room	Audio Conferencing	Video Conferencing	Other
Instructor						
Tutor						
Other (specify)						

Does the instructor check with students at the beginning and during the online course whether they are comfortable using online technologies for the course? (Note: The instructor cannot take for granted that all students who attended the course orientation know how to use all the required technological tasks in the course).

❑   Yes

❑   No

❑   Not applicable

❑   Other (specify)

Does the course provide academic quality such as one would expect in a traditional course?

❑   Yes

❑   No

❑   Not applicable

❑   Other (specify)

If an emergency arises for the instructor, does the course employ a substitute instructor so that students' learning is not interrupted?

❑   Yes

❑   No

❑   Not applicable

❑   Other (specify)

Does the instructor foster an environment of open communication where students can feel comfortable sharing their opinions?

❑   Yes

❑   No

❑   Not applicable
❑   Other (specify)

How soon does the instructor return students' telephone calls?
❑   Within 24 hours
❑   Within 48 hours
❑   Within 72 hours
❑   Within a week
❑   Not applicable
❑   Other (specify)

How soon does the instructor reply to students' e-mail questions?
❑   Within 24 hours
❑   Within 48 hours
❑   Within 72 hours
❑   Within a week
❑   Not applicable
❑   Other (specify)

When e-mail communication is not enough or adequate in clarifying or discussing sensitive issues, are any of the following options available?
❑   Yes
❑   No
❑   Not applicable
❑   Other (specify)

If *yes*, check all that apply:
❑   The instructor calls students
❑   Students can call instructor
❑   Other (specify)
❑   Not applicable

Is the "content expert" who originally designed the course actually teaching the course?

❑   Yes

❑   No

❑   Not applicable

❑   Other (specify)

Is there a "faculty forum" where geographically dispersed faculty members can communicate and exchange ideas to improve their online teaching?

❑   Yes

❑   No

❑   Not applicable

❑   Other (specify)

Is there someone in the institution who compares the accuracy of the course description in the course syllabus with the actual description of the course approved by the curriculum committee?

❑   Yes

❑   No

❑   Not applicable

❑   Other (specify)

Faculty and Support Staff

Check if any of the following have experience teaching online courses.

Role of Individual	Yes	No	NA	Unknown	Other (specify)
Instructor (full-time)					
Instructor (part-time)					
Teaching Assistant					
Tutor					
Technical Support					
Librarian					
Counselor					
Graduate Assistant					
E-Learning Administrator					
Other (specify)					

Check if any of the following have experience as online students themselves.

Role of Individual	Yes	No	NA	Unknown	Other (specify)
Instructor (full-time)					
Instructor (part-time)					
Teaching Assistant					
Tutor					
Technical Support					
Librarian					
Counselor					
Graduate Assistant					
E-Learning Administrator					
Other (specify)					

Check if any of the following receive adequate training to use computer and Internet technologies comfortably to teach/support online courses via any of the following delivery methods.

Role of Individual	Delivery Medium					
	On-line	Face-to-Face	Printed Manual	Training CDs	Blended (Mixture of on-line, f2f, print, CDs, etc.)	Other (specify)
Instructor (full-time)						
Instructor (part-time)						
Teaching Assistant						
Tutor						
Technical Support						
Librarian						
Counselor						
Graduate Assistant						
E-Learning Administrator						
Other (specify)						

Check if any of the following individuals receive training on the *learning management system* (LMS) and/or the *learning content management system* (LCMS) via any of the following delivery methods. (For example, at the University of Maryland University College, students can take the Online Test Drive for WebTycho, the learning management system used in their online courses (http://umuc.edu/distance/de_orien/testdrv_frm.html).

Role of Individual	Delivery Medium										Other (specify)
	On-line		Face-to-Face		Printed Manual		Training CDs		Blended (Mixture of on-line, f2f, print, CDs, etc.)		
	LMS	LCMS	LMS	LCMS	LMS	LCMS	LMS	LCMS	LMS	LCMS	
Instructor (full-time)											
Instructor (part-time)											
Teaching Assistant											
Tutor											
Technical Support											
Graduate Assistant											
Other (specify)											

Check if any of the following individuals receive training on *moderating online discussions* via any of the following delivery methods?

Role of Individual	Delivery Medium					
	On-line	Face-to-Face	Printed Manual	Training CDs	Blended (Mixture of on-line, f2f, print, CDs, etc.)	Other (specify)
Instructor (full-time)						
Instructor (part-time)						
Teaching Assistant						
Tutor						
Graduate Assistant						
Counselor						
Technical Support						
Other (specify)						

Do new faculty (first time teaching online) get technical help to set up their courses?

❑   Yes
❑   No
❑   Not applicable
❑   Other (specify)

Are mentors assigned to new faculty (first time teaching online)?

❑   Yes
❑   No
❑   Not applicable
❑   Other (specify)

Do mentors monitor or observe new faculty throughout the semester and provide feedback for improvement?

❑   Yes
❑   No
❑   Not applicable
❑   Other (specify)

Does the institution have a program or initiative to motivate any of the following instructional and support staff to devote more time and effort in facilitating students' learning? (Note: Online teaching demands more time and effort from instructional staff. Sometimes they have to go out of their way to help learners. If they are not paid for their extra efforts, they may not be motivated to go the extra mile to provide their services.)

❑   Yes
❑   No
❑   Not applicable
❑   Other (specify)

If *yes*, check all that apply:

Instructional and Support Staff	Yes	No	NA	Unknown	Other (specify)
Instructor (full-time)					
Instructor (part-time)					
Teaching Assistant					
Tutor					
Technical Support					
Librarian					
Counselor					
Graduate Assistant					
E-Learning Administrator					
Other (specify)					

Does the instructional staff (e.g., instructor, trainer, and tutor) have adequate computers and connections systems to teach online courses?

❑   Yes

❑   No

❑   Not applicable

❑   Other (specify)

Does the technical support staff have adequate computers and connections systems to support online courses?

❑   Yes

❑   No

❑   Not applicable

❑   Other (specify)

Does the other staff (e.g., librarian, counselor) have adequate computers and connections systems to support online courses?

❑   Yes

❑   No

❑   Not applicable

❑   Other (specify)

In online learning, the instructor plays the roles of a facilitator, mentor and coach. Does the instructor receive training on how to play these roles?

❑   Yes

❑   No

❑   Not applicable
❑   Other (specify)

Does the instructor receive adequate training on the software used in the course in order to answer students' questions or direct them to the appropriate help desk?

❑   Yes
❑   No
❑   Not applicable
❑   Other (specify)

Do technical and other staff receive training on how to communicate with remote learners in difficult situations?

❑   Yes
❑   No
❑   Not applicable
❑   Other (specify)

Does the institution provide any of the following handbooks?

❑   Yes
❑   No
❑   Not applicable
❑   Other (specify)

If *yes*, check all that apply:
❑   Student handbook
❑   Faculty handbook
❑   Tutor handbook
❑   Technical support staff handbook
❑   Other support services staff handbook
❑   E-Learning administrator handbook
❑   Other (specify)
❑   Not applicable

Does the institution provide faculty development courses online? (For example, Capella University provides Faculty Development Courses and Seminars (http://Capella.edu/aspscripts/centers/faculty/resources/fcforms.asp).

❑ Yes
❑ No
❑ Not applicable
❑ Other (specify)

Does the institution provide online seminars for faculty and staff on online learning issues?

❑ Yes
❑ No
❑ Not applicable
❑ Other (specify)

Workload, Class Size, and Compensation

Does the institution have a rewards system to recognize the type of dedication faculty members make in teaching online courses?

❑ Yes
❑ No
❑ Not applicable
❑ Other (specify)

Does the institution provide any kind of compensation, benefits, or rewards to staff who take online training or staff development courses on their own time (not during their work hours)?

❑ Yes
❑ No
❑ Not applicable
❑ Other (specify)

If yes, describe the types of compensations, benefits or rewards.

Does the instructor receive any of the following from the institution as incentives for creating e-learning?

❑   Yes
❑   No
❑   Not applicable
❑   Other (specify)

If *yes*, check the range below:
❑   Release time
❑   Travel funds to conference
❑   Course development funds
❑   Encouragement from peers
❑   Encouragement from administrators
❑   Not applicable
❑   Other (specify)

Does the course limit the number of students per instructor?

❑   Yes
❑   No
❑   Not applicable
❑   Other (specify)

If *yes*, check the range below:

o 10 - 20   o 20 - 30   o 30 - 40   o 40 - 50   o 50 - 60
o 60- 70   o 70 - 80   o 80 - 90   o 90 - 100   o 100 - 110
o No limit   o Not sure   o NA   o Other (specify)

Does the course limit the number of students per tutor?

❑   Yes
❑   No
❑   Not applicable
❑   Other (specify)

If *yes*, check the range below:

o 10 - 20   o 20 - 30   o 30 - 40   o 40 - 50   o 50 - 60
o 60- 70   o 70 - 80   o 80 - 90   o 90 - 100   o 100 - 110
o No limit   o Not sure   o NA   o Other (specify)

Intellectual Property Rights

Does the institution have clear e-learning related policies on the following issues?

❑   Workload and compensation

❑   Intellectual property rights

❑   Not applicable

❑   Other (specify)

Who owns the rights to the materials on the course Website?

❑   Instructor

❑   Content or subject matter expert

❑   Institution

❑   Not applicable

❑   Other (specify)

Do faculty members take course content with them when they leave the institution?

❑   Yes

❑   No

❑   Not applicable

❑   Other (specify)

Pre-Enrollment Services

Are students expected to complete a general and demographic information survey? (Check appropriate option)

❑   Yes, because it is required by institution

❑   Yes, because it is recommended by institution

❑   No

❑   Not applicable

❑   Other (specify)

Does the institution provide contact information of some former students of the course whom new students can contact to get students' perspective of the course? (Note: In traditional on-campus courses students can find former students to get more information about the course and instructor before registration).

❑   Yes

❑   No

❑   Not applicable

❑   Other (specify)

Does the course make previous students' course evaluations available to potential new students?

❑   Yes

❑   No

❑   Not applicable

❑   Other (specify)

Orientation

Check if any of the following individuals are available during online orientation.

Instructional and Support Staff	Yes	No	NA	Other (specify)
Instructor (full-time)				
Instructor (part-time)				
Teaching Assistant				
Tutor				
Technical Support				
Librarian				
Counselor				
Graduate Assistant				
E-Learning Administrator				
Other (specify)				

Do students receive any guidelines on how to interact effectively online?

❑   Yes

❑   No

❑   Not applicable

❑   Other (specify)

Do students receive training on browser skills?

❑   Yes

❑   No

❑   Not applicable

❑   Other (specify)

Are students informed about any of the following technical skills that they can use to become successful in online learning?

❑   Yes

❑   No

❑   Not applicable

❑   Other (specify)

If *yes*, check all that apply:

❑   Students can open both a word processor and browser at the same

❑   Students can take notes in a word processor while using the course browser

❑   Other (specify)

Are there any tips provided during the orientation that can help students to become successful in online learning?

❑   Yes

❑   No

❑   Not applicable

❑   Other (specify)

Does the institution provide a student handbook or other guide that contains important information about relevant institutional policies and procedures?

❑   Yes

❑   No

❑   Not applicable

❑   Other (specify)

If *yes*, check all that apply:
- ❏   It is available online
- ❏   It is mailed to students before orientation
- ❏   It is mailed to students after orientation
- ❏   Other (specify)

Are students required to post their biographies?
- ❏   Yes
- ❏   No
- ❏   Not applicable
- ❏   Other (specify)

If *yes*, can any of the following be included with the biography? Check all that apply:
- ❏   Photographs
- ❏   E-mail addresses
- ❏   Video clips
- ❏   Audio clips
- ❏   Other (specify below)

Does the instructor read and reply to students' biographies when appropriate (as a symbol of a warm welcome)?
- ❏   Yes
- ❏   No
- ❏   Not applicable
- ❏   Other (specify)

Does the course provide a self-test to check students' understanding of orientation and course materials?
- ❏   Yes
- ❏   No
- ❏   Not applicable
- ❏   Other (specify)

Does the institution provide an ID cards to individuals involved in e-learning?

- ❑ Yes
- ❑ No
- ❑ Not applicable
- ❑ Other (specify)

Check if any of the following individuals receive ID cards from the institution.

Role of Individual	Yes	No	NA	Other (specify)
Distance students				
Distance instructors				
Distance Teaching Assistants				
Distance Tutors				
Other (specify)				

Does the institution provide 'I.D. Card Request' form online?

- ❑ Yes
- ❑ No
- ❑ Not applicable
- ❑ Other (specify)

How do distance students and staff apply for ID cards? Check all that apply:

- ❑ Submit their signed and completed I.D. Card Request forms at the campus Registrar's Office where their pictures are taken.
- ❑ They can mail their signed and completed I.D. Card Request forms with their passport sized photos (signed in the back) to the Registrar's Office.
- ❑ They can e-mail their completed I.D. Card Request forms with their digital photos and scanned copy of the drivers' licenses, passport pages (JPEG or GIF file) or any documents showing their identities to the Registrar's Office.
- ❑ Not applicable
- ❑ Other (specify)

Faculty and Staff Directories

Does the institution provide faculty and staff directories online?

- ❑ Yes

❑   No
❑   Not applicable
❑   Other (specify)

If *yes*, check all that are available in the directories:
❑   E-mail address
❑   Telephone number
❑   Fax number
❑   Home page
❑   Mailing address
❑   Other (specify)

Advising

Does the institution provide academic and enrollment advising?
❑   Yes
❑   No
❑   Not applicable
❑   Other (specify)

If *yes*, check all that apply:

Communication Mode	Advising Issues						
	Application Forms	Transcript Review	Course Selection	Transfer Credits	Degree Program Plan	Graduation Forms	Other (specify)
E-mail							
On campus in-person meetings							
Advisors travel to remote sites							
Chat rooms							
Instant messaging							
Bulletin board							
Listserv							
FAQs							
Phone							
Toll-free telephone							
Audio conferencing							
Video conferencing							
Fax							
Regular mail							
Promotional materials							
Admission packets							
Pre-enrollment Materials							
Other (specify)							

Check if any of the following provide academic advising?

Advising by	Yes (check hours)			No	NA	Other (specify)
	Hours Per Week					
	1 - 5	6 - 10	Other			
Instructor (full-time)						
Instructor (part-time)						
Teaching Assistant						
Tutor						
Former Students (Paid)						
Former Students (Volunteer)						
Other (specify)						

Is there a "peer-guide" system available for new students to get guidance from existing students who have completed one or more courses in e-learning?

❑   Yes

❑   No

❑   Not applicable

❑   Other (specify)

Learning Skills Development

Does the course provide a learner's guide?

❑   Yes, available online

❑   Yes, available in print

❑   No

❑   Not applicable

❑   Other (specify)

Services for Students with Disabilities

Check if the institution provides special services for any of the following types of disabilities. Check all that apply:

❑   Visual impairments (blind or impaired sight)

❑   Hearing impairments (deaf or hard of hearing)

❑   Speech impairments

❑   Mobility impairments (restricted manual skills)

- ❑ Dyslexic
- ❑ Mental health difficulties
- ❑ Medical conditions
- ❑ Other learning difficulties (specify)

Bookstore

Does the institution have an online bookstore?
- ❑ Yes
- ❑ No
- ❑ Not applicable

> If *yes*, can students purchase textbooks and packages of course-related supplemental reading materials online from the campus bookstore?
> - ❑ Yes
> - ❑ No
> - ❑ Not applicable
> - ❑ Other (specify)

Does the campus bookstore ship texts and packages of course-related supplemental reading materials to students?
- ❑ Yes
- ❑ No
- ❑ Not applicable
- ❑ Other (specify)

Does the institution provide links to online bookstores where the students can purchase their textbooks?
- ❑ Yes
- ❑ No
- ❑ Not applicable
- ❑ Other (specify)

Does the institution have a developed partnership with an online bookstore where the students can purchase their textbooks with a discount?

❑   Yes

❑   No

❑   Not applicable

❑   Other (specify)

Tutorial Services

Does the institution provide tutorial services?

❑   Yes

❑   No

❑   Not applicable

❑   Other (specify)

Does the instructor (or academic advisor) monitor students' performance during the course?

❑   Yes

❑   No

❑   Not applicable

❑   Other (specify)

> If *yes*, check all that apply:
>
> ❑   Struggling students are advised to use the tutorial services provided by the institution
>
> ❑   Other (specify)

Mediation and Conflict Resolution

Does the institution have a system to accept students' complaints via any of the following?

❑   Yes

❑   No

❑   Not applicable

If *yes*, check all that apply:
- ❑ Telephone
- ❑ Online drop box
- ❑ E-mail
- ❑ Regular mail
- ❑ Other (specify)

Does the institution have an Ombuds Office which students can contact for any of following?
- ❑ Yes
- ❑ No
- ❑ Not applicable

If *yes*, check all that apply:
- ❑ Academic grievances
- ❑ Grade disputes
- ❑ Student/instructor conflicts
- ❑ Sexual harassment
- ❑ Discrimination concerns
- ❑ Other (specify)

Student Newsletter

Does the institution have a student newsletter?
- ❑ Yes
- ❑ No
- ❑ Not applicable
- ❑ Other (specify)

If *yes*, check all that apply:

Newsletter Type	Delivery Format			
	E-mail	Web-based	Print-based	Other (specify)
Daily newsletter				
Weekly newsletter				
Monthly newsletter				
Quarterly newsletter				
Semesterly newsletter				
Other (specify)				

Internships and Employment Services

Check if the institution provides any of the following internship and employment services to students?

Information and Services	Internship	Employment	Other
Information about full-time positions			
Information about part-time positions			
Counseling services			
Other (specify)			

Alumni Affiairs

Does the institution have an alumni association?

❑   Yes

❑   No

❑   Not applicable

❑   Other (specify)

If *yes*, check all that are available:

❑   Alumni Discussion Forum

❑   Directory of Alumni

❑   Other (specify)

Chapter 3

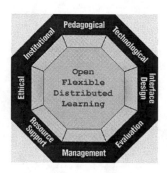

Management Issues

Management of e-learning refers to managing various stages of e-learning processes including; planning, design, production, evaluation, delivery, and maintenance. Simply put, e-learning management needs a system of creating, storing, and maintaining all e-learning contents and resources. Trentin (2003) notes that managing e-learning processes means having a clear idea of complex systems composed of several key elements including content delivery, technology, human resources, and processes of e-learning integration within the organization.

To manage e-learning projects, it is critical to have a clear understanding of the people, process, and products involved in e-learning. Therefore, in this chapter, management of e-learning is discussed in the following order:

1.   People, process, and product continuum in e-learning
2.   Management team
3.   Managing e-learning

Figure 1. E-learning people process product continuum

People, Process and Product (P3) Continuum in E-Learning

In e-learning, people are involved in the process of creating e-learning materials and making them available to its target audience. The PeopleProcessProduct, or the P3 continuum, can be used to map a comprehensive picture of e-learning (Figure 1).

The P3 Model

E-learning process can be divided into two phases: content development and delivery and maintenance (Figure 2). Managing e-learning therefore involves: managing the e-learning content development process (i.e., planning, design, production, and evaluation of e-learning content and resources) and managing delivery and maintenance of e-learning (i.e., implementation of online course offerings, and ongoing updating and monitoring of e-learning environment). Figure 2 provides peopleprocessproduct continuum for content development, and delivery and maintenance of e-learning.

People

Based on the size and scope of the project, the number of individuals involved in various stages of an e-learning project may vary. Some roles and responsibilities may overlap as many e-learning tasks are interrelated and interdependent. A large-sized e-learning project requires the involvement of various individuals. In a small or medium-sized e-learning project, some individuals will be able to perform multiple roles. When an e-learning course is completely designed, developed, taught, and managed by a single individual, it is clear that the same individual performed the role of a content expert, instructional designer, programmer, graphic artist, project manager, and so on. This is an example of a small-size e-learning project. Many of my colleagues have experience develop-

Figure 2. E-learning P3 model

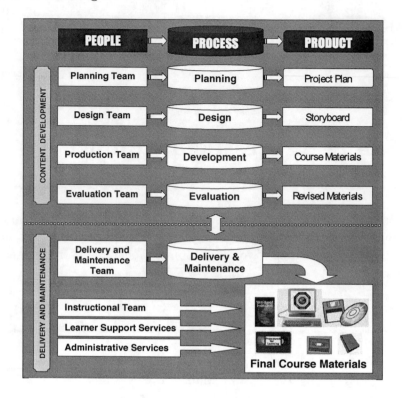

ing their online courses by themselves with intermittent staff support in their institutions. Roles and responsibilities involved in e-learning are described in Table 1.

Process

A typical e-learning process has planning, design, development, evaluation, delivery, and maintenances stages. The e-learning process is iterative in nature. Although the evaluation is a separate stage of the e-learning process shown in Figure 3, an ongoing formative evaluation for improvement (i.e., revision) should always be embedded within each stage of the e-learning process. Individuals involved in various stages of the e-learning process should be in good communication with each other and revise materials whenever needed. The open, flexible, and distributed learning framework (or the OFD framework) discussed in Chapter 1 will help them to make appropriate decisions during various stages of e-learning process.

Table 1. Roles and responsibilities involved in e-learning

Role of Individual	Responsibilities
Director	Directs e-learning initiatives. Develops e-learning plans and strategies.
Project Manager	Supervises the overall e-learning process including design, production, delivery, evaluation, budgeting, staffing, and scheduling. Works with coordinators of various e-learning teams.
Business Developer	Develops business plan, marketing plan, and promotion plan. Coordinates internal and external strategic partnerships.
Consultant / Advisor	Provides independent, expert advice and services during various stages of e-learning.
	Content Development Process
Research and Design Coordinator	Coordinates e-learning research and design processes. Informs management and design teams about the latest data pertaining to online learning activities and research.
Content or Subject Matter Expert	Write course contents and reviews existing course materials (if any) for accuracy and currency.
Instructional Designer	Provides consultation on instructional strategies and techniques for e-learning contents and resources. Helps select delivery format and assessment strategies for e-learning.
Interface Designer	Responsible for site design, navigation, accessibility, and usability testing. Responsible for reviewing interface design and content materials to be compliant with the accessibility guidelines (e.g., section 508 of American disability Act –[ADA]).
Copyright Coordinator	Provides advisement on intellectual property issues relevant to e-learning. Responsible for negotiating permission to use copyrighted materials including articles, books chapters, videos, music, animations, graphics, Web pages, and so on from copyright holders.
Evaluation Specialist	Responsible for evaluation and assessment design and methodology. Conducts and manages student assessment and evaluation of e-learning environments.
Production Coordinator	Coordinates e-learning production process.
Course Integrator	Responsible for getting all pieces of e-learning (e.g., Web pages, chat rooms, Java applets, e-commerce, etc.) working together under a learning management system.
Programmer	Programs e-learning lessons following the storyboard created in the design process.
Editor	Reviews e-learning materials for clarity, consistency of style, grammar, spelling, appropriate references, and copyright information.
Graphic Artist	Uses creativity and style to design graphical images for e-learning lessons.
Multimedia Developer	Responsible for creating multimedia learning objects such as audio, video, 2D/3D animations, simulations, and so on.
Photographer/Videographer (cameraman)	Responsible for photography and video related to e-learning contents.
Learning Objects Specialist	Guides the design, production, and meaningful storage of learning objects by following internationally recognized standards (e.g., SCORM, AICC, IEEE, etc.).
Quality Assurance	Responsible for quality control in e-learning.
Pilot Subjects	Participants in e-learning pilot testing.

Considering the advancement of learning technologies and methodologies, international interoperability standards (e.g., SCORM, AICC, IEEE, etc.), and accessibility guidelines (e.g., section 508 in the USA), technology-based tools can be used for content development and delivery process. For example, for e-learning content development process, a SCORM and 508 compliant content development tool enhanced with instructional design can be developed to assist course designer to create well-designed e-learning contents. For delivery and

Table 1. (continued)

Role of Individual	Responsibilities
Content Delivery and Maintenance Process	
Delivery Coordinator	Coordinates the implementation of e-learning courses and resources.
Systems Administrator	Administers LMS server, user accounts, and network security.
Server/Database Programmer	Responsible for server and database related programming especially for tracking and recording learners' activities.
Online Course Coordinator	Coordinates the instructional and support staff for online courses.
Instructor(or Trainer)	Teaches online courses.
Instructor Assistant	Assists the instructor or trainer in instruction.
Tutor	Assists learners in learning tasks.
Discussion Facilitator or Moderator	Moderates and facilitates online discussions.
Customer Service	Provides generic help and points to appropriate support services based on specific needs of customers (i.e., learners).
Technical Support Specialist	Provides both hardware and software related technical help.
Library Services	Interactive library services for learners who can ask questions to librarians about their research both asynchronous and real time via the Internet.
Counseling Services	Provides guidance on study skills, self-discipline, responsibility for own learning, time management, and stress management, and so on.
Administrative Services	Administrative services include admissions, schedules, and so on.
Registration Services	Responsible for efficient and secure registration process for e-learning.
Marketing	Responsible for marketing e-learning offerings.
Other (specify)	

Figure 3. Interative process of e-learning

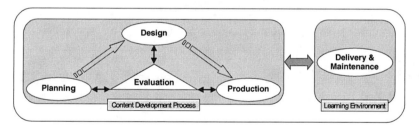

maintenance, a SCORM and 508 compliant learning management system (LMS) can be used (Kang and Khan, in progress, http://BooksToRead.com/omni). Based on an institution's needs and capacity, technology-based tools and systems can be incorporated into the e-learning process to enhance its productivity and delivery. However, tools, and systems alone cannot cover all the activities related to e-learning. Human involvement along with the right tools and systems has the potential for a successful e-learning process. Figure 4 represents the function of content development tool and leaning management system (LMS).

Figure 4. Tools for content development, and delivery and maintenance processes

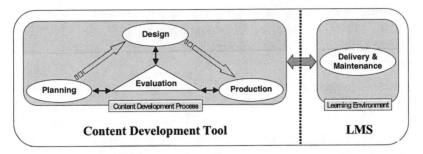

Product

Each stage of the e-learning process has products. Products of an e-learning process may include a project plan, a storyboard, CDROM, printed materials, e-learning materials, and so on (Figure 2). Later in the chapter, products for each stage of the e-learning process are discussed.

Management Team

Managing e-learning projects requires a systemic and integrated team approach. Institutions should establish e-learning management teams with experienced professionals who can maintain quality management systems through continuous monitoring and evaluation to help identify problems in e-learning and suggest solutions (Jung, 2003). The institution should have a competent management team which is responsible for budgeting, staffing, tracking the various e-learning tasks and resources, and the ongoing maintenance of e-learning and blended learning products. The e-learning P3 model (Figure 2) can guide the management in planning, implementing, and operating e-learning projects. Depending on the size and scope of an e-learning initiative, a director or a manager can lead the management team.

Project Manager's Skills

The e-learning project manager's skills should include (not limited to): planning, recruiting, budgeting, supervising, scheduling, assigning tasks to team members, outsourcing projects components, tracking project progress, conducting meetings, presentation, technological, research, interpersonal, oral communication, written communication, consensus building, conflict resolution, and the ability to work with other members on a team.

By combining the expertise of its staff and faculty, Simon Fraser University (SFU) in Canada established an e-learning service, support, and applied research unit called "eLINC" which is run by a management team (http://www.surrey.sfu.ca/elinc).

The management team should have a bird's eye view of its e-learning initiative. E-learning projects should be learner- (or customer-) focused. E-learning projects must produce what customers want. Good products will attract customers. To create well-designed and meaningful e-learning materials and offer it to a globally competitive market, the management should organize its management strategies by following the e-learning P3 model (Figure 2). The P3 model asks: who are involved, what they are supposed to do, and what they produce.

The following is an outline of critical issues the management should consider for initiating the e-learning process:

- Budgeting
- Staffing
- Technology requirements
- Timeline
- Deliverables

Budgeting

E-learning management staff should be knowledgeable about creating and maintaining budgets that reflect the full costs of e-learning projects. As indicated in the *Budgeting and Return on Investment* section of Chapter 2 that e-learning projects need budgets for design and development, delivery, marketing, and ongoing maintenance. Management should be able to communicate effectively about budgets and program needs of the institution. Constant updating and ongoing maintenance are common phenomena for any e-learning projects, therefore it is important to administer the budget efficiently. It means that all e-learning offerings should be up-to-date and well-maintained throughout.

For a sound budget, the management should have a complete understanding of the costs associate with developing and maintaining e-learning. On the Web site titled *"Determining the Cost of Online Courses,"* one can enter data to calculate costs associated with online courses for institutions (http://webpages.marshall.edu/~morgan16/onlinecosts/). In an article entitled "Costs of Developing and Delivering a Web-Based Instruction Course," Harapnuik,

Montgomerie and Torgerson (1998) discuss the primary development and delivery phases of an online course, while also providing a general account of the time spent on each of the phases, and also compares the costs of an online course with those of similar face-to-face courses (http://www.quasar.ualberta.ca/IT/ research/nethowto/costs/costs.html#T5).

Staffing

An e-learning project involves a group of individuals with a wide range of skills, talents, knowledge, experience, and perspectives. They are responsible for doing specific tasks in e-learning. The management should clearly list roles and responsibilities of e-learning team members. Who is doing what? For example, who will conduct pilot testing? Table 1 provides roles and responsibilities generally involved in e-learning. The management should organize these individuals to perform their tasks to achieve project goals. A coordinated and cooperative effort by various individuals will result in the effective management of e-learning projects. For an example, the e-learning management team at the Simon Fraser University clearly lists roles and responsibilities of individuals involved in various e-learning services during the course design, production, and delivery processes (http://www.surrey.sfu.ca/elinc/people.htm).

An e-learning initiative should take full advantage of available resources within its own institution and beyond. The project manager should always be in touch with the institution's information technology and media service department for technical and production assistance. The management should list various e-learning skills acquired by individuals within the institution (Table 2).

After reflecting on various skills acquired by various individuals within the institution, the management will have a comprehensive picture of who should be ideal for what tasks (Table 1) and where outside help is needed. For example, Ms. Lee has skills in instructional design and programming and has been teaching online for the institution (Table 2). The project manager can assign her to tasks related to instructional design, interface design and programming (Table 3).

Table 2. Various e-learning skills acquired by individuals in the institution

Name of the Person	E-Learning Skills
Lee	Taught online courses. Instructional design and programming skills.

Table 3. Person ideal for roles and responsibilities during e-learning process

Name of the Person Ideal for the Job	Director	Project Manager	Business Developer	Consultant / Advisor	Other (specify)	Design Coordinator	Content or Subject Matter Expert	Instructional Designer	Interface Designer	Copyright Coordinator	Evaluation Specialist	Other (specify)	Production Coordinator	Course Integrator	Programmer	Graphic Artist	Multimedia Developer	Editor	Learning Objects Specialist	Quality Assurance	Other (specify)	Evaluation Specialist	Instructional Designer	Interface Designer	Other (e.g., Pilot Subjects)	Delivery Coordinator	Systems Administrator	Server/Database Programmer	Other (specify)
Lee							√	√						√															

Table 4. Outsourcing e-learning

Name of the Outside contractor Ideal for the Job	Director	Project Manager	Business Developer	Consultant / Advisor	Other (specify)	Design Coordinator	Content or Subject Matter Expert	Instructional Designer	Interface Designer	Copyright Coordinator	Evaluation Specialist	Other (specify)	Production Coordinator	Course Integrator	Programmer	Graphic Artist	Multimedia Developer	Editor	Learning Objects Specialist	Quality Assurance	Other (specify)	Evaluation Specialist	Instructional Designer	Interface Designer	Other (e.g., Pilot Subjects)	Delivery Coordinator	Systems Administrator	Server/Database Programmer	Other (specify)
XYZ																√	√												

Based on what is appropriate for the institution, the management can either hire new people or outsource to external sources (Table 4). For example, XYZ is a multimedia and graphic design firm with several years of experience in e-learning. The management can outsource all of its multimedia and graphic design to XYZ.

Some team members may not be located at the same place or even know or meet each other. The project manager should make sure each member of the team accomplishes his or her part of the tasks so the entire e-learning project runs smoothly. It is important for all team members to understand his/her responsibilities. They should have strong commitment for the projects. Strong cross-communication skills are needed for individuals involved in e-learning. All members must have patience as continually emerging issues may demand new changes and modification in e-learning, which in turn can be more work for all members. Therefore, ability, and willingness to work on teams is one of the most critical characteristics of an e-learning team member.

Managing a successful e-learning project involves good communications among all parties involved in the e-learning process. The management should develop a communication and knowledge sharing system so that all individuals involved in e-learning can benefit for shared knowledge. While developing the project plan during the planning stage (Figure 2) of e-learning process, the management should consider establishing an e-learning project support site and a knowledge management site to support communication and knowledge sharing process (discussed later in the chapter).

Technology Requirements

The management should make sure that all e-learning materials are developed following technology requirements stated by the institution. For example, if the required modem speed for learners is 28.8 KPS, then all multimedia learning materials should be developed with keeping the institution's technology requirement in mind. Technology requirements are discussed in detail in the *Hardware* and *Software* sections of Chapter 4.

Timeline

In e-learning, each task in the process is a part of the whole project. E-learning content development is a collaborative process where tasks are interdependent. The management should develop a timeline that lists the start and end dates for all tasks. For example, the evaluation specialist will start the development of test items for Lesson 1 on May 12 and finish on May 26 (Table 5). All members should be sensitive to the project timeline for their respective tasks. Since each task is interdependent, failure to follow the dates in the timeline will delay the project.

Deliverables

In each stage of the e-learning process, individuals are involved in producing one or more learning products. These products are deliverables to their respective clients as schedule deadlines. For example, the product of a design process is a storyboard. At the end of the design process, the design team must give the storyboard (i.e., deliverable) to the production team. Deliverables in e-learning

Table 5. E-learning project timeline

Activity	Project Manager	Business Developer	Consultant / Advisor	Design Coordinator	Content or Subject Matter Expert	Instructional Designer	Interface Designer	Copyright Coordinator	Evaluation Specialist	Production Coordinator	Course Integrator	Programmer	Graphic Artist	Multimedia Developer	Editor	Learning Objects Specialist	Quality Assurance	Delivery Coordinator	Systems Administrator	Server/Database Programmer	Other (specify)	Start Date	End Date	Comments
Test items for lesson 1									√													May 12	May 26	Same Tests formats will be used in other lessons.

can include (but not limited to): contents, storyboards, HTML files, XML files, PDF files, word processing files, audio clips, video clips, graphics, animations, CDROMs, DVDs, and so on.

Managing E-Learning

Managing e-learning environment involves the management of two major e-learning phases: managing e-learning content development process (i.e., planning, design, production, and evaluation of e-learning contents and resources) and managing the e-learning environment (i.e., delivery of online course offerings and ongoing updating and monitoring of e-learning environment). In this section, the following are discussed:

1.   Managing content development process
2.   Managing e-learning environment

Managing Content Development Process

Managing content development process includes assigning responsibilities to individuals during various stages of the process and supervising the entire development process. The following is an outline of this section:

Figure 5. Content migration

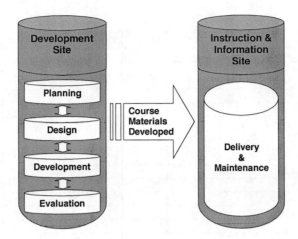

- Development and instruction site
- Roles and responsibilities
- Development process

Development and Instruction Site

At the beginning of the content development process, it is important to plan for storage of various e-learning materials. E-learning projects can be organized and hosted on two different servers or sites: development site and instruction and information site. Once course materials are completed, then they can be migrated from the development site to instruction and information site (Figure 5).

Development Site

This is an internal production site. Individuals involved in the content development process will be accessing this site with their usernames and passwords. Completed content materials are transferred to the instructional and information site. At the end of an e-learning content development process, the development site can be discontinued.

Table 6. Roles of individual during various stages of e-learning

Role of Individual	E-Learning Stages				
	Planning	Design	Production	Evaluation	Delivery
Director					
Project Manager					
Business Developer					
Consultant / Advisor					
Other (specify)					
Research and Design Coordinator					
Content or Subject Matter Expert					
Instructional Designer					
Interface Designer	√	√	√	√	
Copyright Coordinator					
Other (specify)					
Production Coordinator					
Course Integrator					
Programmer					
Graphic Artist					
Multimedia Developer					
Photographer/Videographer (cameraman)					
Editor					
Learning Objects Specialist					
Quality Assurance					
Pilot Subjects					
Other (specify)					
Evaluation Specialist					
Other (specify)					
Delivery Coordinator					
Systems Administrator					
Server/Database Programmer					
Other (specify)					

Instruction and Information Site

E-learning instruction and information site contains courses and information. This is an external site where visitors can get information about the institution, program, and courses. This site is open to visitors. However, only registered students will be able to access their courses with their usernames and passwords. Visitors can review demo courses. Managing an instructional site includes the coordination of services of instructional, administrative, and support services staff.

Roles and Responsibilities

There are many roles and responsibilities involved in e-learning process. In Table 6, roles and responsibilities are categorized under planning, design, production, evaluation, and delivery. As indicated earlier in the chapter that some roles and responsibilities of individuals may overlap as many e-learning tasks are interrelated and interdependent. Some of these roles may be needed in more than one stage of the process. The project manager should assign individuals with tasks that they can perform in various stages of the project whenever needed. In some instances, the project manager may ask individuals to assume multiple roles. For example, instructional designers can assist during the planning, design, production, and evaluation stages of e-learning. After identifying institution's expertise (Table 2), the management should creatively assign individuals to appropriate tasks in one or more stages of the process.

Development Process

As indicated earlier, a typical e-learning content development process involves planning, design, development, and evaluation stages (Figure 3). The e-learning process is iterative in nature. Although the evaluation is shown as a separate stage of the e-learning process, but an ongoing formative evaluation for improvement (i.e., revision) should always be embedded within each stage of the e-learning process. Individuals involved in various stages of the e-learning process should be in good communication with each other and revise materials whenever needed. Figure 6 highlights the PeopleProcessProduct Continuum for Content Development Process.

Planning Stage

At the planning stage, the planning team should develop a project plan by analyzing various aspects of people, process, and products involved in e-learning initiative. This plan must be pedagogically and financially sound and should guide each e-learning team (e.g., production, evaluation, delivery, maintenance, instructional, and support services) to engage in their respective assigned activities. During the planning stage, the team creates a project plan that clearly identifies the people, process, and product of each phase of e-learning process including, design, development, evaluation, delivery, and maintenance. The plan also indicates the estimated completion time for each task.

Director, project manager, research and design coordinator, and instructional designer can work together to make the project plan pedagogically sound. The

Figure 6. People-process-product continuum for content development process

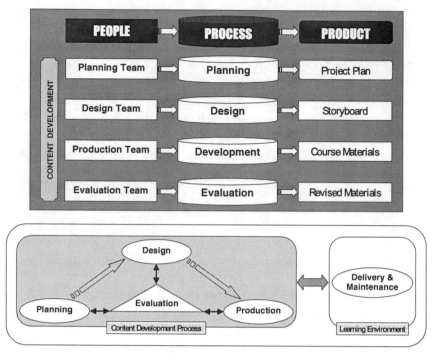

business developer can work with the team to make sure the plan is financially sound. The business developer should also develop a marketing plan by including empirical data on institution's course quality, student satisfaction, and retention. Research and design coordinator can assist the business developer in analyzing student data and can also provide valuable research information about e-learning on a regular basis. Outside consultants can also be part of the planning process.

The *product* of the e-learning planning process is a sound e-learning *project plan* (see Figure 6). The e-learning project plan provides guidance during the various stages of the e-learning process. E-learning designers, developers, evaluators, instructional, and support staff should follow the guidelines of the e-learning project plan to provide a meaningful learning environment for learners.

Design Stage

At the design stage, the research and design (R&D) coordinator leads the e-learning course design process. With a comprehensive understanding of learners' needs, institutional capabilities, and the experience in e-learning design and research, the R&D coordinator is responsible for reviewing course content for pedagogical soundness and the selection of appropriate delivery medium. In the design stage, instructional designers work with subject matter experts, interface designers, copyright coordinators, and evaluation specialists (Table 1).

The *product* of a course design process is the *storyboard* (Figure 6). During the development of the course storyboard, the design team communicates with the production and delivery teams for any technical and production related issues.

Instructional designers are knowledgeable about how to use various attributes and resources of the Internet and digital technologies to design e-learning activities. Based on the content types, they can incorporate instructional strategies and techniques best suited for target audience (see *methods* and *strategies* section in Chapter 5). However, it is important to note that instructional design services may not be adequately available in some institutions. With the knowledge of instructional design, instructors can design their own online courses.

In developing course contents, subject matter experts may decide to use copyrighted materials from outside sources. Instructional designers can work with copyright coordinators for negotiating permission to use copyrighted materials.

Interface designers are responsible for the presentation of e-learning content. They make sure that the learners can use e-learning materials in a userfriendly environment. They work with instructional designers to create a consistent look and feel for all e-learning contents.

Following the overall goals and objectives of the course, evaluation specialists can design various assessment strategies to measure students' performance in e-learning. Please note that an instructional designer with skills and knowledge in interface design and assessment strategies may be able to play the role of an interface designer and evaluation specialist in small-to-medium sized e-learning projects. However, for largesized e-learning projects, services of interface designers and evaluation specialists are very critical.

Many online universities use outside content experts for their courses. I provided the content for a course entitled "Developing and Implementing e-Learning Systems" for the masters in education in e-learning at the Jones International University (JIU), the first fully online, accredited university in the USA. As a content author, I worked directly with an instructional designer who guided me in the preparation of the course content. While writing content for the course, I had to provide a list of all additional readings (e.g., journal articles and books chapters). The copyright acquisition coordinator at JIU negotiated with publishers about using their materials in the course. However, one publisher's price was unacceptable by JIU. Therefore, I had to revise the content to select another reading assignment. Information about the course is available at: http://jiu-web-a.jonesinternational.edu/eprise/main/JIU/courses/displayCourse?banner= &leftnav=&navdir=&backtext=&dirhome= &distributor=&COURSENO =EDU754.

Production Stage

At the course production stage, the production team creates the online course from the course storyboard developed during the design phase. The production coordinator leads e-learning production process. Team members include, but are not limited to; course integrator, programmer, graphic artist, multimedia developer, photographer/videographer (cameraman), editor, learning objects specialist, and quality assurance (see Table 1 for Roles and Responsibilities).

The production coordinator should make sure the timeline is maintained for all deliverables. The e-learning production process is time consuming. It is a collaborative process where each member does his or her own specific tasks for a course. For example, the course integrator cannot put all parts of a lesson together if each member does not provide his or her part of the task on time.

In e-learning, members of the development team can be remotely located. The production coordinator should make sure members are in good communication with each other and in compliance with due dates for their respective tasks. Members should put their works in designated areas on a centralized server (which we can call a "development server"). The development server becomes a collaborative workspace for the e-learning members. Recently, I served as a consultant for an e-learning project development at the World Bank where I worked with the project manager, instructional designers, graphic artists, and programmers who were remotely located.

It is always wise to create two resource sites during the content development process: project support site or PSS and knowledge management or KM site. The PSS can be discontinued after the project is completed whereas the KM site is an ongoing knowledge sharing site for an organization.

- **Project Support Site for Content Development:** During the e-learning content development process, a project support site can be established to assist members of various e-learning teams to access and share project related information. Project support site provides a one-stop-shopping node for projects notes, proposals, design standards, graphics, media files, documents, or any other materials that assist the production team in creating e-learning materials for the instructional site (Gotcher, 2001).

- **Knowledge Management Site:** Institutions should create a knowledge management (KM) system, which is an ongoing process of identifying, creating, disseminating, and utilizing knowledge to serve the individuals within their own institutions and beyond. Everyone in an institution has something unique to contribute and the knowledge management (KM) system should develop an environment where an individual's knowledge is valued and rewarded, and the free flow of ideas is encouraged. Individuals must feel proud to share their knowledge and benefit from collective wisdom of others. A well-designed KM system means everybody wins. However, if an individual in a workplace does not have job security, he/she may not volunteer to share his or her unique knowledge to the institution's KM.

Rosenberg (2001) notes, "Knowledge management supports the creation, archiving, and sharing of valuable information, expertise, and insight within and across communities of people and organizations with similar interest and needs" (p. 66). Wagner (2001) says that regardless of how KM is implemented in individual organizations, it provides organizations with an essential source of competitive advantage in the information economy by capturing, storing, and making accessible its full array of intellectual assets. To improve the existing knowledge base and elimination of redundancies, institutions must constantly update their KM systems.

Santosus and Surmacz (2001) discuss how KM can be used in golf:

Think of a golf caddie as a simplified example of a knowledge worker. Good caddies do more than carry clubs and track down wayward balls. When asked, a good caddie will give advice to golfers, such as: "The wind makes the ninth hole play 15 yards longer." Accurate advice may lead to a bigger tip at the end of the day. On the flip side, the golfer, having derived a benefit from the caddie's advice, may be more likely to play that course again. If a good caddie is willing to share what he knows with other caddies, then they all may eventually earn bigger tips. How would KM work to make this happen? The caddie master may decide to reward caddies for sharing their tips by offering them credits for pro shop merchandise. Once the best advice is collected, the course manager would publish the information in notebooks (or make it available on PDAs [Personal Digital Assistants]), and distribute them to all the caddies. The end result of a well-designed KM program is that everyone wins. In this case, caddies get bigger tips and deals on merchandise, golfers play better because they benefit from the collective experience of caddies, and the course owners win because better scores lead to more repeat business.

Table 7. Existing and needed contents for knowledge management

List Contents Needed for KM	Status of Contents		Contributor Name	Require Permission?		
	Need to Create	*Already Exist*		*Yes*	*No*	*NA*

Similarly, the KM idea can be used in e-learning. For example, in an e-learning course's discussion forum, learners can list the most important skills they found useful to become successful online learners. Learners' input saved from various semesters can be archived and made accessible to learners. After reviewing learners' input, an instructor can develop a "Top 10 List of Things that are Useful for Online Learning" that can be useful to all learners. To design a KM, it is necessary to list what content exists, what content must be created and what content requires permission (Table 7).

Once the course is created, it is important to do a pilot testing with a representative group of diverse learners. For the pilot testing, learners can access the course at the development server with a password. These learners can be remotely located. For an efficient evaluation of the pilot project, the course should be designed to receive learners' comments on a specific page. For an example, if a learner finds a symbol is culturally offensive on a page, he or she can attach his or her comments with that specific graphic or the page. The production coordinator can make these comments available to responsible team members. Data from the pilot testing will provide valuable information about what works and what does not work. Instructional designer and the interface designers can work with the production team and revise the course whenever appropriate.

The *product* of the production process is *course materials* ready for pilot testing (see Figure 6).

Evaluation Stage

Several phases of evaluation can be conducted during an e-learning process. These evaluations are conducted to improve the effectiveness of e-learning materials. There are two types of evaluation: formative and summative. By conducting ongoing formative evaluation, we can improve the e-learning product as it is being developed. Summative evaluation is usually conducted to do the final assessment of the e-learning products. However, e-learning projects undergo ongoing evaluation for improvement. Therefore, formative evaluation is inherent to e-learning development process.

Driscoll (1998) discussed the four evaluation phases of Web-based training: subject-matter expert, rapid prototype, alpha class, and beta class. Alpha and beta tests are terms usually used by software development companies to evaluate the effectiveness of software. Development of e-learning and supplemental materials can use these four types of formative evaluation strategies. However, the budget and timeline may influence the extent to which an e-learning project undergoes these evaluation phases.

Subject-matter expert (SME) review can be conducted to examine the accuracy and currency of content. More than one SME can review the course content. Experts can provide information about whether the content is complete and accurate, and follows a logical sequence based on prerequisites. They can also review any graphics and multimedia used in the course in order to provide feedback on their relevancy to the course content.

Rapid prototype evaluations can be conducted to determine the effectiveness of the overall design of a course by only examining a sample module of that course. Prototypes can range from paper-based lesson to an online module. For an online module, prototype target subjects can be a representative sample with diverse background (e.g., familiarity with content and Internet technologies) from different locations. Evaluation specialists, design teams, and production teams can analyze the prototype data.

Based on the rapid-prototype evaluation, the production team can make changes to the course and conduct an alpha class evaluation to test the effectiveness of the entire course. The production coordinator or a production team member can act as an instructor for this fully developed course piloted on the Internet. Based on his or her experience as instructor, the developer can make changes to the course. If an e-learning course does not require an instructor, then beta-class evaluation is not required. Chapter 7 discusses usability testing of course materials in detail.

Beta class evaluation is done to test the effectiveness of directions to the instructor. How well the course runs when taught by someone rather than the developer (Driscoll, 1998). Beta evaluation is usually done by an instructor who was not involved in the design and development of the course. The course is offered to learners and the context that resemble the intended learners and environment.

Instructional designer and interface designers can assist evaluation specialist in analyzing learners' feedback from the pilot testing. With learners' feedback, the evaluation specialist communicates with design and production teams for revising the course accordingly.

Once the course is delivered, students' evaluation (both ongoing and at the end of the course) will provide important feedback, which can be used by the design and development team to revise the course materials for improvement. However, some institutions may not have the luxury of having non-students to pilot test their courses. In such situations, the method of testing on the first delivery on real students can be useful.

The *product* of the evaluation process is *revised course materials* which can be offered to learners (Figure 6). The following summary of evaluation phases is taken with permission from Table 102 (p. 220) of Driscoll's 1999 book entitled "Web-Based Training."

Summary of Evaluation Phases

	Purpose	Evaluators	Methods
Subject-Matter Expert Review	Check the accuracy and completeness of the content.	SMEs in the content and topic area.	• Document review • Interview
Rapid Prototype	Check effectiveness of the instructional design, clarity of directions, and ease of interactions, using a single lesson.	Representative learners of high, moderate, and low skill levels.	• One-on-one meetings • Interviews • Observations
Alpha Class	Test the effectiveness of the complete course being taught as a stand-alone program or as a facilitated program taught by the developer.	Representative learners of high, moderate, and low skill levels.	• Interviews • Surveys • Observations
Beta Class	Assess the effectiveness of the complete course and clarity and usefulness of directions for instructor.	Instructors and representative learners of high, moderate, and low skill levels.	• Interviews • Surveys • Observations

Managing E-Learning Environment

Once the course materials are developed and approved by the institution, the next step is to make them available to target audience. Therefore, managing an e-learning environment includes the delivery and maintenance of all learning materials (Figure 7).

The following will be discussed in this section of managing e-learning environment:

1.   E-Learning delivery and maintenance
2.   Course offerings

Figure 7. People-process-product continuum for delivery and maintenance process

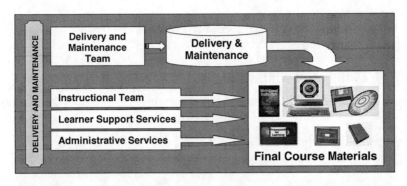

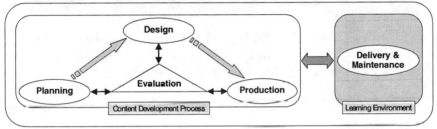

E-Learning Delivery and Maintenance

All online course materials should be accessible by the learners at anytime from anywhere in the world. All supplemental course materials (e.g., CD, DVD, audio, and video cassette, book, course pack, etc.) should be delivered to learners. The delivery and maintenance (D&M) team should maintain an effective and efficient learning environment with their assigned roles and responsibilities (Table 1).

The D&M team maintains the leaning management system (LMS) and data-bases, provides technical support to students, instructors, and support staff, and manages LMS user accounts and network security. They also provide technical assistance to the design and production teams in the areas of software and hardware related issues for e-learning. They are also responsible for duplicating and distributing learning materials and installing and maintaining the course.

In some institutions, e-learning materials are developed by outside vendors who may be responsible for migrating all learning materials to the institution's server. Overall, the D&M team is responsible for ongoing updating and monitoring of e-learning environment.

An example of how an e-learning delivery team functions, visit Simon Fraser University's site at http://www.surrey.sfu.ca/elinc/deliveryteam.htm.

The *product* of the delivery and maintenance process is well-maintained *course materials* available for registration (Figure 7).

Ongoing Updating and Monitoring of E-Learning Environment

Ongoing updating and monitoring is a major part of e-learning maintenance process. Individuals involved in maintenance should keep e-learning materials updated on a regular basis. They should also check if all links and resources are active.

Security Measures

Security measures includes access control and information confidentiality. No institutions are immune from hackers. Academic networks can be targets of hackers if they lack security. The Associated Press reported on March 7, 2003 that hackers broke into a University of Texas database and stole the names, social security numbers, and e-mail addresses of more than 55,000 students, former students, and employees (http://www.washingtonpost.com/wp-dyn/articles/A53517-2003Mar6.html).

Course Offerings

In this section, course offerings are discussed in terms of all e-learning and blended courses offered through academic, corporate, vendor, or independent education and training providers. When a student takes an online course, what does he or she expect from the instructor, technical support, library, and so on? Whatever instructional, administrative, and learner support services he or she might need — may come from various sources at the course, program, and institution levels. The extent to which support services are available may be dependent on which division of an institution is offering the course. Is it offered through the distance education office, continuing education office, academic departments or programs, training department, or human resources office within the institution? Is the institution an academic, a corporate, or a vendor? Or is the course offered by an independent individual? Based on who is offering the course, the management of instructional, administrative and learner support services may vary from context to context. In this section, the following are discussed:

- Distribution of information

- Instruction stage

Distribution of Information

Information distribution covers the delivery of both online and off-line e-learning materials including schedule, syllabus, announcements, course relevant contact information, learning and testing materials, and students' grades from quizzes, assignments, exams, and projects. Students can have access to testing materials and their grades by entering their password. It is important to note that students with disabilities may not be able to use online course materials if they are not designed appropriately for them.

Courses can provide some course content to be downloaded in .wp, .doc, pdf file formats. However, different countries may use different versions of word processing software, therefore it is always advisable to have .pdf option available or other common format (HTML, text).

E-Learning Instruction Stage

At the course instruction stage (Figure 8), instructional and support services staff are involved (Table 7). The instructional and support services staff may include (but not limited to): instructor, tutor, course facilitator, discussion moderator, technical support, librarian, counselor, customer service, registration and administrative staff, and so on.

When a course is offered, the ISS is at the front line. Students deal with ISS. They expect uninterrupted and meaningful learning environments. The online course coordinator should make sure that registered students receive orientation for the course and ISS support is available as promised. The course coordinator should always be in touch with the delivery and maintenance team to resolve any technical problems that the ISS team may encounter during the course.

Depending on the organizational structure of an institution, the course coordinator may have to work with various departments within the institution including registration, admissions, legal offices, and so on. It is important to note that the instructional staff for online courses may or may not be part of the e-learning management team, they may be managed by academic or training departments. Table 8 can be used to assign responsibilities for individuals in ISS team. For example, Mr. Lee has been teaching online courses for several years and has experience in providing technical support. With Mr. Lee's instructional and technical skills, he can be assigned the responsibilities of an instructor, discussion facilitator/moderator, and technical support for course number 201.

Figure 8. E-learning environment

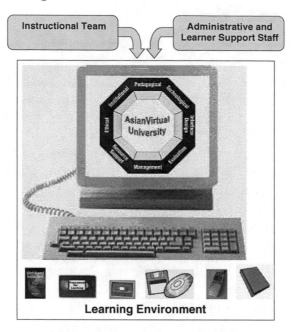

Table 8. Person ideal for roles and responsibilities during instructional process

Name of the Person Ideal for the Job	Administrative						Instructional								Learner Support						
	Project Manager	Admission	Registration	Payment	Bookstore	Financial Aid	*Online Course Coordinator*	Instructor (or Trainer)	Instructor Assistant	Tutor	Discussion Facilitat or/Moderator	Learning Objects Specialist	Copyright Coordinator	Guest Speaker (or outside Expert)	*Delivery Coordinator*	Systems Administrator	Server/Database Programmer	Customer Service	Technical Support Specialist	Library Services	Counseling Services
Mr. Lee (Course Number 201)								√			√								√		

Question to Consider

Can you think of any e-learning management issues not covered in this chapter?

Activity

1. Using Internet search engines, locate at least one article covering any of the following management issues, and analyze the article from the perspective of its usefulness in e-learning:

    *   Budgeting

    *   Staffing

    *   Technology requirement

    *   Timeline

    *   Deliverables

    *   Project support site for content development

    *   Knowledge management site

    *   Ongoing updating and monitoring of e-learning environment

    *   Security measures

2. Locate an online program and review its e-learning management issues using the relevant management checklist items at the end of Chapter 3.

References

Driscoll, M. (1998). *Web-based training*. San Francisco, CA: Jossey-Bass Pfeiffer.

Gotcher, L.T. (2001). Project support sites: A project management tool for constructing Web-based training. In B.H. Khan (Ed.), *Web-based training*, (pp. 321-324). Englewood Cliffs, NJ: Educational Technology Publications.

Harapnuik, D., Montgomerie, T.C. & Torgerson, C. (1998). Costs of developing and delivering a Web-based instruction course. *Proceedings of WebNet*

98—World Conference of the WWW, Internet, and Intranet. Association for the Advancement of Computing in Education. Charlottesville.

Jung, I. (2003). Online education for adult learners in South Korea. *Educational Technology, 43*(3), 9-16.

Rosenberg, M.J. (2001). *E-learning: Strategies for delivering knowledge in the digital age.* New York: McGraw-Hill.

Santosus, M. & Surmacz, J. (2001). The ABCs of knowledge management. *CIO Magazine. http://www.cio.com/forums/knowledge/edit/kmabcs.htm*

Trentin, G. (in press). Managing the complexity of e-learning systems. *Educational Technology, 43*(6), 36-42.

Wagner, E.D. (2001). Emerging learning trends and the World Wide Web. In B.H. Khan (Ed.), *Web-based training,* (pp. 33-50). Englewood Cliffs, NJ: Educational Technology Publications.

Management Checklist

People

Underline the roles and responsibilities of individuals listed in the "Responsibilities" column of the table below. Also, add any additional duties in the blank spaces.

Role of Individual	Responsibilities
Director	Directs e-learning initiatives. Develops e-learning plans and strategies.
Project Manager	Supervises the overall e-learning process including: design, production, delivery, evaluation, budgeting, staffing and scheduling. Works with coordinators of various e-learning teams.
Business Developer	Develops business plan, marketing plan, and promotion plan. Coordinates internal and external strategic partnerships.
Consultant / Advisor	Provides independent, expert advice and services during various stages of e-learning.
	Content Development Process
Research and Design Coordinator	Coordinates e-learning research and design processes. Informs management and design teams about the latest data pertaining to online learning activities and research.
Content or Subject Matter Expert	Writes course content and reviews existing course materials (if any) for accuracy and currency.
Instructional Designer	Provides consultation on instructional strategies and techniques for e-learning contents and resources. Helps select delivery format and assessment strategies for e-learning.
Interface Designer	Responsible for site design, navigation, accessibility and usability testing. Responsible for reviewing interface design and content materials to be compliant with the accessibility guidelines (e.g., section 508 of American disability Act - ADA).
Copyright Coordinator	Provides advice on intellectual property issues relevant to e-learning. Responsible for negotiating permission to use copyrighted materials including articles, books chapters, videos, music, animations, graphics, Web pages, etc. from copyright holders.
Evaluation Specialist	Responsible for evaluation and assessment design and methodology. Conducts and manages student assessment and evaluation of e-learning environments.
Production Coordinator	Coordinates e-learning production process.

(continued from previous page)

Role of Individual	Responsibilities
Course Integrator	Responsible for getting all pieces of e-learning (e.g., Web pages, chat rooms, Java applets, e-commerce, etc.) working together under a learning management system.
Programmer	Programs e-learning lessons following the storyboard created in the design process.
Editor	Reviews e-learning materials for clarity, consistency of style, grammar, spelling, appropriate references and copyright information.
Graphic Artist	Uses creativity and style to design graphical images for e-learning lessons.
Multimedia Developer	Responsible for creating multimedia learning objects such as audio, video, 2D/3D animations, simulations, etc.
Photographer/ Videographer (cameraman)	Responsible for photography and video related to e-learning contents.
Learning Objects Specialist	Guides the design, production and meaningful storage of learning objects by following internationally recognized standards (e.g., SCORM, AICC, IEEE, etc.).
Quality Assurance	Responsible for quality control in e-learning.
Pilot Subjects	Participants in e-learning pilot testing.
Content Delivery and Maintenance Process	
Delivery Coordinator	Coordinates the implementation of e-learning courses and resources.
Systems Administrator	Administers LMS server, user accounts and network security.
Server/Database Programmer	Responsible for server and database related programming especially for tracking and recording learners' activities.
Online Course Coordinator	Coordinates the instructional and support staff for online courses.
Instructor (or Trainer)	Teaches online courses.

(continued from previous page)

Role of Individual	Responsibilities
Instructor Assistant	Assists the instructor or trainer in instruction.
Tutor	Assists learners in learning tasks.
Discussion Facilitator or Moderator	Moderates and facilitates online discussions.
Customer Service	Provides generic help and points to appropriate support services based on specific needs of customers (i.e., learners).
Technical Support Specialist	Provides both hardware and software related technical help.
Library Services	Interactive library services for learners who can ask questions to librarians about their research both asynchronously and in real time via the Internet.
Counseling Services	Provides guidance on study skills, self-discipline, responsibility for own learning, time management and stress management, etc.
Administrative Services	Administrative services include admissions, schedules, etc.
Registration Services	Responsible for efficient and secure registration process for e-learning.
Marketing	Responsible for marketing e-learning offerings.
Other (specify)	

Management Team

Check the skill level of the management team. (check all that apply):

Skill Types	Management Team														
	Director					Project Manager					Other (specify)				
	Excellent	Good	Fair	Poor	NA	Excellent	Good	Fair	Poor	NA	Excellent	Good	Fair	Poor	NA
Recruiting															
Supervising															
Budgeting															
Planning															
Scheduling															
Assigning tasks to team members															
Interpersonal															
Presentation															

(continued from previous page)

Skill Types	Management Team														
	Director					Project Manager					Other (specify)				
	Excellent	Good	Fair	Poor	NA	Excellent	Good	Fair	Poor	NA	Excellent	Good	Fair	Poor	NA
Technological															
Research															
Outsourcing projects components															
Tracking Project Progress															
Conducting Meetings															
Oral Communication															
Written Communication															
Consensus Building															
Conflict Resolution															
Ability to work with others on a team															
Other (specify)															
Other (specify)															
Other (specify)															

Are the budgets maintained efficiently to keep e-learning updated and running without any financial problems?

❑   Yes

❑   No

❑   Not applicable

List various e-learning skills acquired by individuals within the institution (for an example, Mr. Lee's skills are listed in the table below):

Name of the Person	E-Learning Skills
Lee	Taught online courses. Instructional design and programming skills.

Identify person ideal for roles and responsibilities during e-learning process. Check all that apply:

Name	Director	Project Manager	Business Developer	Consultant / Advisor	Other (specify)	Design Coordinator	Content or Subject Matter Expert	Instructional Designer	Interface Designer	Copyright Coordinator	Evaluation Specialist	Other (specify)	Production Coordinator	Course Integrator	Programmer	Graphic Artist	Multimedia Developer	Editor	Learning Objects Specialist	Quality Assurance	Other (specify)	Evaluation Specialist	Instructional Designer	Interface Designer	Other (e.g., Pilot Subjects)	Delivery Coordinator	Systems Administrator	Server/Database Programmer	Other (specify)

Identify outside contractor (i.e., outsourcing) ideal for roles and responsibilities during e-learning process. Check all that apply:

Name	Director	Project Manager	Business Developer	Consultant / Advisor	Other (specify)	*Design Coordinator*	Content or Subject Matter Expert	Instructional Designer	Interface Designer	Copyright Coordinator	Evaluation Specialist	Other (specify)	*Production Coordinator*	Course Integrator	Programmer	Graphic Artist	Multimedia Developer	Editor	Learning Objects Specialist	Quality Assurance	Other (specify)	Evaluation Specialist	Instructional Designer	Interface Designer	Other (e.g., Pilot Subjects)	*Delivery Coordinator*	Systems Administrator	Server/Database Programmer	Other (specify)

Are all e-learning materials created based on the institution's stated technology requirements?

❑   Yes

❑   No

❑   Not applicable

❑   Other (specify)

Indicate timeline for project activities and responsible individuals (for example, Mrs. Smith is responsible for developing test items for lesson 1):

Activity	Project Manager	Business Developer	Consultant / Advisor	Design Coordinator	Content or Subject Matter Expert	Instructional Designer	Interface Designer	Copyright Coordinator	Evaluation Specialist	Production Coordinator	Course Integrator	Programmer	Graphic Artist	Multimedia Developer	Editor	Learning Objects Specialist	Quality Assurance	Delivery Coordinator	Systems Administrator	Server/Database Programmer	Other (specify)	Start Date	End Date	
Test items for lesson 1									Smith													May 12	May 26	Same Test format will be used in other lessons.

Managing Content Development Process

Check the roles and responsibilities of individuals listed below during various stages of e-learning. (Note: some individuals may assume multiple roles.) Check all that apply:

Role of Individual	E-Learning Stages				
	Planning	Design	Production	Evaluation	Delivery
Director					
Project Manager					
Business Developer					
Consultant / Advisor					
Other (specify)					
Research and Design Coordinator					
Content or Subject Matter Expert					
Instructional Designer					
Interface Designer					
Copyright Coordinator					
Other (specify)					
Production Coordinator					
Course Integrator					
Programmer					
Graphic Artist					
Multimedia Developer					
Photographer/Videographer (cameraman)					
Editor					
Learning Objects Specialist					
Quality Assurance					
Pilot Subjects					
Other (specify)					
Evaluation Specialist					
Other (specify)					
Delivery Coordinator					
Systems Administrator					
Server/Database Programmer					
Other (specify					

Is there a project support site (PSS) for e-learning design, development, evaluation and delivery teams?

❑   Yes

❑   No

❑   Not applicable

If *yes*, check if the PSS has any of the following. (check all that apply):

❑   File upload feature to exchange documents by team members

❑   Project progress

- ❑ Start and end dates for each task
- ❑ The project delivery date (e.g., course is completed)
- ❑ Document updates
- ❑ Updates of new resources
- ❑ Minutes from facetoface meetings
- ❑ Minutes from online conferencing
- ❑ Team members' e-mail addresses
- ❑ Team members' phone numbers
- ❑ Team members' mobile phone numbers
- ❑ Team members' fax numbers
- ❑ Project due dates
- ❑ Meeting information (date, time and place)
- ❑ Graphics relevant to the project
- ❑ Audio files relevant to the project
- ❑ Video files relevant to the project
- ❑ Other (specify)

Is there a knowledge management (KM) site?

- ❑ Yes
- ❑ No
- ❑ Not applicable
- ❑ Other (specify)

List the content, contributor and copyright information for the knowledge management site:

List Contents Needed for KM	Status of Contents		Contributor Name	Require Permission?		
	Need to Create	Already Exist		Yes	No	NA

Does the institution acquire permission to use copyrighted materials for its knowledge management (KM) system from the individual copyright holders who work in the institution?

❑   Yes

❑   No

❑   Not applicable

❑   Other

Does the institution acquire permission to use copyrighted materials for its knowledge management (KM) system from copyright holders outside the institution?

❑   Yes

❑   No

❑   Not applicable

❑   Other

Does the knowledge management (KM) system have ongoing review processes to amend, delete and update its information?

❑   Yes

❑   No

❑   Not applicable

❑   Other

Managing Delivery and Maintenance

Does the course provide test make-ups for students who get disconnected from the course Website during the test?

❑   Yes

❑   No

❑   Not applicable

❑   Other (please describe below)

Do students get notified when course Websites are not available, for example, down for maintenance or upgrades?

❑   Yes

❑   No

❑   Not applicable

Are the course materials updated regularly (e.g., are Web pages maintained, up to date, etc.)?

❑   Yes

❑   No

❑   Not applicable

Is the date of the revision or update being displayed prominently?

❑   Yes

❑   No

❑   Not applicable

Does the course inform students who is responsible for updates?

❑   Yes

❑   No

❑   Not applicable

Is there a link to send comments and suggestions for the Website or course?

❑   Yes

❑   No

❑   Not applicable

Check if students are notified about any changes in due dates or other course relevant matters (e.g., if the server hosting the course goes down) via any of the following. Check all that apply:

❑   E-mail

❑   Announcement page

❑   Alert boxes

❑   Running footer added to a page

  ❑   Phone call
  ❑   Mail
  ❑   Other (specify)
  ❑   Not applicable

Check if any of the following security measures are implemented in the course. Check all that apply:
  ❑   Login with password
  ❑   Digital signature
  ❑   Firewall
  ❑   Randomization of test questions to prevent sharing of answers
  ❑   Other (specify)

Does the course have encryption (i.e., a secure coding system) available for students to send confidential information over the Internet?
  ❑   Yes
  ❑   No
  ❑   Not applicable
  ❑   Other (specify)

Does the course have encryption (i.e., a secure coding system) available for online payment?
  ❑   Yes
  ❑   No
  ❑   Not applicable
  ❑   Other (specify)

Is this course password protected so that only enrolled students have access to this course?
  ❑   Yes
  ❑   No
  ❑   Not applicable

Does the course provide students with designated and secure (e.g., password protected) online spaces to store their personal notes and resources?

❑   Yes

❑   No

❑   Not applicable

Does the course have archives of previous students' discussion forum transcripts on topical issues?

❑   Yes

❑   No

❑   Not applicable

Can a hacker change contents of the course Web pages?

❑   Yes

❑   No

❑   Not applicable

❑   Not sure

Can outsiders crash the online course?

❑   Yes

❑   No

❑   Not applicable

❑   Not sure

Does the course protect students' information from outsiders (hackers)?

❑   Yes

❑   No

❑   Not applicable

❑   Other (specify)

Are unregistered individuals given access to any part of the course?

❑   Yes

❑   No

❑   Not applicable

If *yes*, list types of contents/materials that unregistered individuals can have access to:

Is there any other reliable way to submit assignments for an online class?
❑ Yes
❑ No
❑ Not applicable

If *yes*, check all that apply:
❑ Students can send assignments on disks
❑ Students can send hard copies of assignments
❑ Students can provide the addresses of their personal Websites where their assignments or projects are located
❑ Other (specify)

Does the course have a system of keeping records of student interactions? (Note: This is a privacy issue. Students' permission may be needed to use their postings).
❑ Yes
❑ No
❑ Not applicable

If *yes*, check all types of interactions:
❑ Between students
❑ Between students and instructor(s)
❑ N/A
❑ Other (specify below)

Which division of the academic institution, corporation or vendor does offer the course? Check all that apply:

Settings	Course Offered By						
	Distance Education Office	Continuing Education	Academic Department	Training Department	Human Resources Office	Individual Faculty / Trainer	Other
Academic							
Corporate							
Vendor							
Other							

Does the course have the space to store student projects and products?

❑   Yes

❑   No

❑   Not applicable

❑   Other (specify)

Does the course allow students to print out the online contents of the Web pages? (Note: This may be useful for students who prefer reading them off-line. However, sometimes, unnecessary blank pages are printed in addition to actual Web pages. Course designers should minimize this problem by providing special tips to users to avoid printing blank pages.)

❑   Yes

❑   No

❑   Not applicable

Does the course have page counters? (Note: Page counters are useful for students to keep track of where they are in relation to the lesson. For example, 1 of 5 pages.)

❑   Yes

❑   No

❑   Not applicable

Check if any of the following supplemental materials are used in the course. Check all that apply:

❑   Books

❑   e-books

❑   Videotape

- ❑   Audiotape
- ❑   CD-ROM
- ❑   Printed packet
- ❑   Other
- ❑   Not applicable

Check if the course provides a brief biography of any of the following. Check all that apply:

Role of Individual	Yes	No	NA	Other (specify)
Instructor				
Tutor				
SME or Content Expert				
Technical Support				
Library Support				
Counselor				
Other				

Does the course provide back-up materials or alternative activities for students (i.e., what will students do?) if any of the following is either not operating properly or unavailable during a scheduled lesson period?

- ❑   Yes
- ❑   No
- ❑   Not applicable

If *yes*, check all that apply:
- ❑   Access to the courseware
- ❑   Discussion forum
- ❑   Chat room
- ❑   E-mail and mailing list
- ❑   Books
- ❑   E-books
- ❑   Online resources
- ❑   Library materials
- ❑   Study guide
- ❑   Instructor
- ❑   Tutor

❑   Technical support
❑   Other (specify)

Does the course syllabus provide any of the following options? Check all that apply:
❑   Course description and overview
❑   Course goals/objectives
❑   Course calendar
❑   Instructor's synchronous office hours
❑   Instructor's contact information
❑   Technical support staff's contact information
❑   Technical support staff's synchronous office hours
❑   Schedule of readings
❑   Assignments/projects information
❑   Assignments/projects due dates
❑   Attendance policy
❑   Late assignment policy
❑   Online discussion participation requirement policy
❑   Academic dishonesty policy
❑   Exams administration
❑   Grades
❑   Technology requirements
❑   Required textbook
❑   Recommended texts
❑   E-books
❑   Course relevant resources (on the Web)
❑   Other (specify)

Does the course indicate whether the course content is best viewed by a specific browser?
❑   Yes
❑   No
❑   Not applicable

Does the course indicate whether the course content is best viewed by a specific monitor display (e.g., 800, 1024, etc.)?

❑   Yes

❑   No

❑   Not applicable

If *yes*, please indicate the screen size:

Does the course provide a class distribution list (list containing student e-mail addresses) for students?

❑   Yes

❑   No

❑   Not applicable

Does the course provide a list containing students' telephone numbers to be used by other students in the course? (Note: Students permission may be needed to make their telephone numbers available to other students.)

❑   Yes

❑   No

❑   Not applicable

Does the course provide a list containing students' addresses? (Note: Students permission may be needed to make their addresses available to others.)

❑   Yes

❑   No

❑   Not applicable

Does the course provide the option for students to create their personal Web pages?

❑   Yes

❑   No

❑   Not applicable

Identify the ideal person for the roles and responsibilities required during the instructional stage of e-learning process. Check all that apply:

Name of the Person Ideal for the Job	Administrative							Instructional									Learner Support							
	Project Manager	Admission	Registration	Payment	Bookstore	Financial Aid		Online Course Coordinator	Instructor (or Trainer)	Instructor Assistant	Tutor	Discussion Facilitator/Moderator	Learning Objects Specialist	Copyright Coordinator	Guest Speaker (or outside Expert)		Delivery Coordinator	Systems Administrator	Server/Database Programmer	Customer Service	Technical Support Specialist	Library Services	Counseling Services	

Does the course have a system of keeping track of student submissions, online quizzes, etc.?

❑   Yes

❑   No

❑   Not applicable

How often do the instructor or technical support staff check the accessibility of the course Website?

❑   Daily

❑   Weekly

❑   Monthly

❑   Other (specify)

Does the course track attendance in the discussion forum?

❑   Yes

❑   No

❑   Not applicable

Does the course have a system of reminding students about upcoming assignments?

❑   Yes
❑   No
❑   Not applicable

    If *yes*, how are students reminded? Check all that apply:
    ❑   E-mail
    ❑   Phone
    ❑   Announcement on the course Website
    ❑   Other (specify)

Does the instructor acknowledge receipt of assignments within:

❑   24 hours of initial receipt
❑   48 hours of initial receipt
❑   72 hours of initial receipt
❑   Other
❑   Not applicable

Does the course keep a computer log with data about learners' participation in online discussions?

❑   Yes
❑   No
❑   Not applicable

    If *yes*, does the log data include any of the following? Check all that apply:
    ❑   Number of posts
    ❑   Time spent on each discussion topic
    ❑   Other (specify)

When does the instructor return students' assignment with feedback and grade?

❑   Within 7 days of initial receipt
❑   Within 10 days of initial receipt

❑   Within 14 days of initial receipt

❑   Other (specify)

❑   Not applicable

Is each student's progress monitored regularly?

❑   Yes

❑   No

❑   Not applicable

If *yes*, how (describe below)?

Are students contacted if their assignments are not received on time?

❑   Yes

❑   No

❑   Not applicable

Does the course have a private space for student interaction (for example a student "lounge" or "cafe" where there is no faculty surveillance?)

❑   Yes

❑   No

❑   Not applicable

Does the course have an automatic response mechanism which can send confirmation of receipt of assignments or other submissions immediately?

❑   Yes

❑   No

❑   Not applicable

Does this course provide a direct link to send messages for help if students are having problems?

❑   Yes

❑   No

❑   Not applicable

Can learners (or participants) link to outside Websites (as references) from their postings on the course discussion forum? (Note: Online articles or documents relevant to discussion topics can enhance the quality and the validity of postings.)

❑   Yes

❑   No

❑   Not applicable

Does the course allow students to do online editing of their already submitted (existing) documents? (Note: This may be useful for students who want to edit their already published materials for the Web-based courses.)

❑   Yes

❑   No

❑   Not applicable

Does the course allow students to upload their documents (or files) to the course Website?

❑   Yes

❑   No

❑   Not applicable

Does the course Website have an option for students to submit their assignments online?

❑   Yes

❑   No

❑   Not applicable

Does the course allow students to leave or broadcast messages for the entire class, cohort, group or program (bulletin board, listserv, etc.)?

❑   Yes

❑   No

❑   Not applicable

Does the course provide a place for groups to work on documents?

❑   Yes

❑   No

❑   Not applicable

Does the course allow students to replace an existing document by simply uploading the new document? (Note: This may be useful for students who want to replace an already published document with new changes.)

❑   Yes

❑   No

❑   Not applicable

Does the course have a mechanism to help students convert and upload students' presentation slides which are created using presentation software such as PowerPoint?

❑   Yes

❑   No

❑   Not applicable

Chapter 4

Technological Issues

The technological dimension of the framework examines issues of technology infrastructure in e-learning environments. This includes infrastructure planning, hardware, and software. The following is an outline of this chapter:

- Infrastructure planning
- Hardware
- Software

Infrastructure Planning

An e-learning environment is built on digital infrastructure. Boettcher and Kumar (2000) noted that, as with our physical infrastructure, this digital infrastructure needs to be designed, planned, built, maintained, and staffed. Rosenberg (2001) states that e-learning infrastructure is based on an institution's technological capabilities to deliver and manage e-learning. Therefore, it should be based on an architecture that depends on open, published standards, reusability of components, serviceability, and maintainability.

A stable, long-lived, and widely available infrastructure is highly desirable in e-learning. Boettcher and Kumar (2000) stress it clearly that an e-learning infrastructure should be:

- Scalable (i.e., able to handle growth in terms of increased number of users, more demanding applications, and a greater variety of applications.),
- Sustainable (i.e., resilient and pliable enough to survive and accommodate technology changes as well as the test of time),
- Reliable, and
- Consistently available (i.e., 24 hours a day, seven days a week).

Infrastructure planning for e-learning, therefore, should focus on issues such as (but not limited to):

- What technological and technical capabilities are required to support e-learning;
- What essential skills (i.e., digital literacy) are needed by learners, instructors, and support staff to be successful in ever changing digital learning environment;
- What standards and guidelines should be followed to create and share learning contents; and
- What policies should be employed for technology infrastructure.

Technological and Technical Capabilities

An institution should have reliable and efficient networks and competent technical staff to support e-learning. All stakeholder groups (i.e., students, faculty members, support staff, etc.) should be informed about the technological and technical capabilities of the institution's e-learning program. They should be well-informed about any latest news and updates about the networks and their accessibility. They should be given appropriate orientation about any updates in the system that may affect their work.

From time to time, there may be some unexpected problems associated with networks. For an example, servers hosting online courses can go down at the most unexpected and inconvenient times. Therefore, the course designer(s) should consider ways to minimize the crashed servers' effect by including alternative learning activities.

Also, there might be situations where learners can get kicked off in the middle of a test or an assignment. One of my students who took an online course at her work was kicked off the Internet after two hours. In one of the courses, she got kicked out in the middle of a test and she had to start over again — did not save the work she had done. It is not clear whether getting kicked of is planned ("after two hours") or inadvertent ("middle of a test"). This can relate to network's connection or to service provider or equipment failure.

Students new to e-learning may be anxious when they encounter network and server problems. Both technical and instructional staff must work together to come up with innovative solutions to tackle these problems. To help students troubleshooting and reduce anxiety, Spitzer (2001) recommends setting up a "buddy system" so that students will have at least one person whom they can call upon for assistance.

Digital Literacy

In the digital learning environment, all stakeholder groups (i.e., students, faculty members, support staff, etc.) should be digital literate in order for them to participate actively in e-learning. Digital literacy may include (but is not limited to) skills needed to use browser, search engines, file transfer, scanner, digital cameras, and familiarities with terms and jargons, and so on. Branigan (2002) notes: "New technologies — like computers and the Internet — require different skill sets, such as reading graphs, searching databases, and thinking critically. How do different search engines work? How does one tell good information from bad?"

Shareable and Reusable Learning Objects

In designing e-learning courses, we have to organize and manage course content in the most effective and efficient ways possible. These e-learning course contents are comprised of a myriad of learning objects (LOs), also known as content objects, knowledge objects, reusable information objects, and reusable learning objects.

What is a learning object? Simply, a learning content that supports learning. A learning object is a discrete piece of instructional content that meets a learning objective (Masie, 2002). Wagner (2001) notes that learning objects are stand-alone pieces of information that are reusable in multiple contexts depending on the needs of the individual user. A learning object can be any entity, digital, or non-digital, that can be used or referenced in e-learning. A learning object can be physical, such as text, a workbook, or a CD-ROM, or online, such as electronic text, a GIF graphic image, a QuickTime movie, or a Java applet (McGreal & Roberts, 2001).

Wiley (2000) defines a learning object as "any digital resource that can be reused to support learning." Wiley's definition includes anything that can be delivered across the network on demand, be it large or small. Examples of smaller reusable digital resources include digital images or photos, live data feeds (like stock tickers), live or prerecorded video or audio snippets, small bits of text, animations, and smaller Web-delivered applications, like a Java calculator. Examples of larger reusable digital resources include entire Web pages that combine text, images, and other media or applications to deliver complete experiences, such as a complete instructional event.

Why should e-learning institutions and e-learning designers pay attention to LOs? The simple answer is: LOs are reusable and shareable. According to Singh (2001), "Learning objects simplify collaboration, sharing, and reuse of instructional content, meeting the needs for rapid creation of new content while easing the burden of capturing knowledge as objects from subject matter experts" (cited in Merkow, 2001). Learning objects are an important part of e-learning contents, and we should create them by following guidelines provided by international standardization bodies, including the Institute for Electrical and Electronic Engineers (IEEE). The United States Defense Department has been developing interoperability standards for e-learning products known as Sharable Content Object Reference Model (SCORM). SCORM is a set of interrelated technical specifications built upon the work of the AICC (Aviation Industry CBT Committee), IMS (Instructional Management Systems), and IEEE to create one unified content model. These specifications enable the reuse of Web-based learning content across multiple environments and products (http://xml.coverpages.org/scorm.html). If we can create LOs following the interoperability standards, we can save money. We can reuse them whenever we need them and then share them with others.

In order to create cost effective e-learning courses, institutions should start identifying and compiling common learning objects (LO) that are used in their courses. The compiled LOs can then be put at a server that can be accessed by course designers. Having these LOs centrally available, course designers can efficiently identify, locate, and reuse them in multiple courses. Learners can also use them in their projects. These LOs can, therefore, be used whenever and wherever appropriate by individuals within the institution and beyond. These reusable LOs will save both money and energy.

If more and more institutions start to meaningfully archive LOs with proper names with standard metadata tags[1], e-learning resources will increasingly improve. Not every single course has to create same LOs over and over again. Instructional designers or individuals involved in designing e-learning should meta-tagged LOs with appropriate information such as author, language, date, instructional objectives, and so on. Once all the LOs are meta-tagged, e-learning designers can efficiently find, select, retrieve, combine, use/re-use, and target them for appropriate use (Masie, 2002).

The Digital Library Network for Engineering and Technology or DLNET provides a Learning Object (LO) Packaging Tool that assists users in the content submission process. DLNET describes the LO Packaging Tool as below:

The learning objects created in your submission process become a resource that can be used by learners, educators, and course builders. The most important attribute of a learning object is that it encapsulates metadata (information about the resource and its creator) together with the resource you contributed. Thus, wherever your learning object goes, so will the metadata that you created, ensuring that your contribution is recognized and acknowledged by all who use it. (URL: http://www.dlnet.vt.edu/ ToolGuidelines.jsp)

Followings are some resources for Learning Objects:

* Learning Objects Spark an E-learning Revolution, by Mark S. Merkow (http://www.techlearning.com/db_area/archives/WCE/archives/ mmerkow.htm).
* The Web site at http://reusability.org/read hosts chapters from an online version of *The Instructional Use of Learning Objects* book where authors connect learning objects to instruction and learning.

Policy

E-learning institutions should develop policies and guidelines on various technology related issues for hardware, software, and networks. Do employees have permission to install additional software in their office desktop/laptop computers? This is a software policy related issue. In e-learning courses, sometimes additional software (e.g., plug-ins) will be required. For example, when an e-learning course uses an external resource site, the course designers have no control over any changes, which may be made by that Web site. The external site may require users to install additional software or may require permission to access. Institution should have clear policies about such situations. Otherwise, individuals (i.e., faculty members, employees, and students using institutional computers) involved in e-learning may suffer. Institution should also have clear policies on who can or cannot have access to the learning objects available at the institution.

Hillsborough Community College (HCC) established its network policy and guidelines that serve as the standard of conduct when using HCC's Internet access. Violation of any of these provisions may result in the immediate, temporary, or permanent termination of access to the network as well as subjected to other penalties and disciplinary action (URL: http://www.hcc.cc.fl.us/webadmin/policies). Davidson College has developed hardware and software policies for faculty, students, and staff for academic, non-commercial use of the Information Technology Services (URL: http://www2.davidson.edu/administration/hr/hr_empguide/collsrvcs_its.asp).

Hardware

Hardware for e-learning may include computers, servers, modems, networking devices, wireless devices, printers, scanners, cameras, microphone, storage devices (e.g., hard drives, CD-ROM, DVD, etc.), and other equipments.

Software

Software for e-learning may include (but not limited to) word processors, e-mail packages, presentation programs, graphic software, reader software, browsers and plug-ins, spreadsheets, databases, learning management systems (LMS)[2], learning content management systems (LCMS), authoring tools and enterprise software[3], and so on. E-learning designers use many of these types of software in creating and maintaining e-learning programs. However, students only use LMS, word processor and plug-ins to participate in an e-learning course.

For e-learning designers, it is critical to know the functionality of LMS and LCMS. For graphical representations of how an e-learning course is built from a repository of learning objects and the functions of an LMS and LCMS, visit http://64.224.94.100/itimegroup/lcms/. Stacey (2001) discusses the differences between LMS and LCMS functionality as in Table 1.

Considering the open, flexible and distributed nature of e-learning (see *Blended Strategies* section in Chapter 5), it seems that there is a tremendous need for SCORM and W3C (Web Content Accessibility Guidelines) compliant authoring tools/LMSs that provide instructional design support for content development and delivery by addressing issues encompassing the eight dimensions of the e-learning framework. The framework can provide diagnostic checklist which can be used to ensure that no important factor is omitted from the design of e-learning, whatever its scope or complexity. Dr. Myunghee Kang of Ewha Womans University in South Korea and I have been involved in the development of *OMNI*, a SCORM and W3C compliant content development tool and learning management system with instructional design support (see Figure 1). For more information about OMNI initiative, visit http://www.BooksToRead.com/omni.

For an e-learning initiative to be viable in an institution, it should be meaningfully integrated into a central system of the organization which is shared by various systems or departments, including finance, procurement, customer relationship management (CRM), enterprise resource planning (ERP), human resources management system (HRMS), and so on. To run an effective and efficient e-learning initiative, an institution should make sure that whatever LMS, LCMS, or authoring tools they use for e-learning can be easily integrated into its enterprise-level systems (CRM, ERP, HRMS).

An enterprise portal through centralized communication and collaboration can bring together the different communities that make up an organization and reinforce the sense that each has an important role and stake in the company (Voth, 2002). Pepperdine University's Graziadio School of Business uses an enterprise portal, called GraziadioNet, to promote a sense of community within its seven centers throughout California, allowing far-flung students to feel a part of a single school.

Table 1. LMS and LCMS functionality

LMS Functionality	LCMS Functionality
• Schedules and registers learners into online and off-line courses • Keeps learner profile data • Launches e-learning courses • Tracks learner progress through courses • Manages classroom based learning • Provides learning administrators with the ability to manage learning resources including labs and classrooms (resource management) • Supports learner collaboration • Automates use of competency maps to define career development and performance paths (skills gap analysis) • Creation of test questions and administration of test • Performance reporting learning results • Interconnectivity with Virtual Classroom (VC), LCMS and enterprise applications	• Content migration and management • Content creation tools • Workflow tools to manage content development process • Learning object repository • Organizing reusable content • Content reuse and adaptive individualized learning paths based on learning objects • Asynchronous collaborative learning including discussion groups • Testing and certification • Reporting of results • Delivering content in multiple formats (online, print, PDA, CD-ROM, etc.) • Providing content navigational controls (look and feel) • Interconnectivity with Virtual Classroom, LMS and enterprise applications

Figure 1. Conceptual model for content development and learning management system

Questions to Consider

Can you think of any specific issues not covered in this chapter for technological issues of e-learning?

Can you explore the different technologies that are currently being used to support learning in your own institution?

Activity

1. Using Internet search engines, locate at least one article relevant in any one of the following technological topics, and analyze the article from the perspective of its usefulness in e-learning:

    *   Learning objects
    *   Learning management systems
    *   Learning content management systems
    *   Enterprise software

2. Using Internet search engines, locate at least one article relevant to any one of the following technological topics, and analyze the article from the perspective of its usefulness in e-learning:

    *   Learning objects
    *   Learning management systems
    *   Learning content management systems
    *   Enterprise software

3.   Locate an online program and review its technological issues using the relevant technological checklist items at the end of Chapter 4.

References

Boettcher, J.V. & Kumar, M.S. (2000). The other infrastructure: Distance education's digital plant. *Syllabus, 13*(10).

Branigan, C. (2002). Digital literacy is essential for students. *Assistant eSchool News,* from *http://eschoolnews.com/news/showStory.cfm?ArticleID= 3592*

Masie, E. (2002). *Making sense of learning specifications and standards: A decision maker's guide to their adoption.* Saratoga Springs, NY:The Masie Center.

McGreal, R. & Roberts, T. (2001). A primer on metadata for learning objects: Fostering an interoperable environment. *E-Learning.* Retrieved January 24, 2003, from *http://www.elearningmag.com/elearning/article/ articleDetail.jsp?id=2031*

Merkow, M.S. (2001). Learning objects spark an e-learning revolution, from *http://www.techlearning.com/db_area/archives/WCE/archives/ mmerkow.htm*

Rosenberg, M.J. (2001). *E-learning: Strategies for delivering knowledge in the digital age.* New York: McGraw-Hill.

Spitzer, D.R. (2001). Don't forget the high-touch with the high tech in distance learning. *Educational Technology, 41*(2), 51-55.

Stacey, P. (2001). Learning Management Systems (LMS) & Learning Content Management Systems (LCMS) - E-Learning an Enterprise Application? Retrieved January 24, 2003, from *http://www.bctechnology.com/statics/ pstacey-oct2601.html*

Voth, D. (2002). Why enterprise portals are the next big thing. *E-Learning, 3*(9), 25-29. Retrieved January 24, 2003, from *http:// www.elearningmag.com/elearning/article/articleDetail.jsp ?id=36877*

Wagner, E.D. (2001). Emerging learning trends and the World Wide Web. In B.H. Khan (Ed.), *Web-based training.* Englewood Cliffs, NJ: Educational Technology Publications.

Wiley, D.A. (2000). Connecting learning objects to instructional design theory: A definition, a metaphor, and a taxonomy. In D.A. Wiley (Ed.), *The instructional use of learning objects: Online version.* Retrieved January 24, 2003, from *http://reusability.org/read/chapters/wiley.doc*

Endnotes

[1]   Tags affixed to learning objects to explain what's inside: content, objectives, author, language, date, version, level, assessment, and more (source: http://www.internettime.com/itimegroup/lcms/index.htm).

[2]   An LMS registers the learner and records completion whereas an LCMS assembles learning objects into learning paths personalized to the needs of the individual learner (source: http://www.internettime.com/itimegroup/lcms/index.htm).

[3]   Enterprise software is integrated software that facilitates the flow of information among all the processes of an organization's supply chain (source: http://sap.mis.cmich.edu/sap.intro/lesson01/sld006.htm).

Technological Checklist

Infrastructure Planning

Does the institution have a technology plan that clearly describes the process of acquiring, maintaining, and upgrading hardware and software required for e-learning?

❏   Yes

❏   No

❏   Not applicable

❏   Other (specify)

Check if the institution's network system has any of the following characteristics of a stable, long-lived, and widely available technology infrastructure? (check all that apply):

❏   Scalable

❏   Sustainable

❏   Reliable

❏   Consistently available

❏   Other (specify)

Does the course have orientation programs that provide technical training to students before starting the course?

❑   Yes

❑   No

❑   Not applicable

❑   Other (specify)

Does the institution have personnel who can assist learners in setting up their computers before starting the course?

❑   Yes

❑   No

❑   Not applicable

Is the course Website hosted on the course provider's own system?

❑   Yes

❑   No

❑   Not applicable

If *no*, check all that apply

❑   Hosted on a commercial system (with monthly fee)

❑   Hosted on an outside system (free of charge)

❑   Other (specify)

Does the course provide the following information about the institution's network system to learners?

❑   Bandwidth capacity

❑   Limitations of its networks

❑   Not applicable

❑   Other (specify)

Is there a "buddy system" established in the course so that learners will have at least one person who they can call to do some preliminary troubleshooting or just ask advice?

❑   Yes

❑   No
❑   Other (specify)

What happens to a pre-scheduled exam or chat when the server is down?
❑   Exam or chat is rescheduled
❑   Exam or chat is postponed until the server is up and running
❑   Exam is done offline
❑   Not applicable
❑   Not sure
❑   Other(specify)

How efficient was the course server in offering access to the course Web pages?
❑   Very efficient
❑   Efficient
❑   Fair
❑   Good
❑   Poor
❑   Not applicable
❑   Other (specify)

Does the course provide alternative off-line learning activities if the course server goes down?
❑   Yes
❑   No

Does the course provide toll-free numbers where students can dial to connect to the Internet/Web free of charge?
❑   Yes
❑   No
❑   Not applicable
❑   Other (specify)

Does the course provide a list of Internet Service Providers (ISPs) with which learners reported having encountered problems in accessing and using the course Website? (Note: An institution cannot recommend, endorse, or promote any specific ISP best suited to the course requirement. Therefore, it cannot provide a list of Internet Service Providers best suited to the course requirements.)

❑   Yes

❑   No

❑   Not applicable

❑   Other (specify)

Does the course provide e-mail accounts to students?

❑   Yes

❑   No

❑   Not applicable

If *yes*, specify the storage space or disk quota per student:

If some students do not have enough computer expertise or skills to participate in the course, does the course offer any training sessions or direct students to appropriate resources so that they can get the necessary skills to fully participate in the various activities of the course?

❑   Yes

❑   No

❑   Not applicable

❑   Other (specify)

Is a learner's full participation in the course tied to accessing the technological components at specific times? (Note: If the course requires students to participate in synchronous activities, there will be designated times when students need to be at their course workstations. If the workstation is at home and the student is at his/her office during those times, this becomes an issue. In some circumstances, it may be cost effective for students to use PDAs, tablet PCs or other devices. However, students should check whether PDAs and other devices can use the course if it is only designed for regular PCs.)

❑ Yes

❑ No

❑ Not applicable

Are the minimum capabilities (e.g., browser, software compatibility, data transfer speeds) for an adequate Internet service provider specified in the course?

❑ Yes

❑ No

❑ Not applicable

Is there any financial aid available for students to purchase the necessary technology required for the course?

❑ Yes

❑ No

❑ Not applicable

Check if any of the following individuals have any of the following digital literacy skills. Check all that apply:

Role of Individuals	Digital Technology Skills																							
	Browser			Search Engines			File Transfer (ftp)			Scanner			Digital Camera			Creating CDs			Terms and Jargon			Other		
	Yes	No	NA	Yes	No	NA	Yes	No	NA	Yes	No	NA	Yes	No	NA	Yes	No	NA	Yes	No	NA	Yes	No	NA
Learner																								
Instructor (full-time)																								
Instructor (part-time)																								
Trainer																								
Trainer Assistant																								
Tutor																								
Technical Support																								
Help Desk																								
Librarian																								
Counselor																								
Graduate Assistant																								
Administrator																								
Other (specify)																								

Has the institution created any reusable and shareable learning objects or LOs (i.e., smallest pieces of learning contents)?

❑   Yes

❑   No

❑   Not applicable

> If *yes*, check all that apply
>
> ❑   Individuals within the institution can use without permission (free of charge)
>
> ❑   Individuals within the institution can use with permission (free of charge)
>
> ❑   Individuals outside the institution can use without permission (free of charge)
>
> ❑   Individuals outside the institution can use with permission (free of charge)
>
> ❑   Individuals outside the institution can use (with fees)
>
> ❑   Other (specify)

Are learning objects created following international interoperability standards?

❑   Yes

❑   No

❑   Not applicable

> If *yes*, specify the standards (e.g. SCORM).

If appropriate, are all learning objects available in the course reusable?

❑   Yes

❑   No

❑   Not applicable

Is there a search facility to search for various learning objects within the institution?

- ❑ Yes
- ❑ No
- ❑ Not applicable

Can learning objects available in the institution be used by its own students for their projects?

- ❑ Yes
- ❑ No
- ❑ Not applicable

   If *yes*, check all that apply:
   - ❑ Can use them without the permission of the institution
   - ❑ Cannot use them without the permission of the institution
   - ❑ Can use them with a fee
   - ❑ Other
   - ❑ Not applicable

Can learning objects available in the institution be used by outsiders?

- ❑ Yes
- ❑ No
- ❑ Not applicable

   If *yes*, check all that apply:
   - ❑ Anyone can use without the permission from the institution
   - ❑ Cannot use without the permission of the institution
   - ❑ Can use with a fee
   - ❑ Other
   - ❑ Not applicable

Is the cost of required hardware, software and the types of Internet connection (e.g., T1, DSL, cable modem, etc.) a deterrent to taking this course?

❏   Yes

❏   No

❏   Not applicable

Does the institution have special arrangements with vendors to offer students special prices for hardware and/or software?

❏   Yes

❏   No

❏   Not applicable

If *yes*, list below:

Hardware and Software	Vendor Name	Price

Are any disk quotas allocated for students in their accounts on the institution's server?

❏   Yes

❏   No

❏   Not applicable

If *yes*, can student request for increased disk quotas for special projects?

❏   Yes

❏   No

❏   Other (specify)

Are students given specific guidelines on how much computer expertise they need to have to participate in the course? (For example, a list of things they should know how to do on the Internet.)

❑   Yes

❑   No

❑   Not applicable

Are there time limits for how long learners can be logged on to the course?

❑   Yes

❑   No

❑   Not applicable

❑   Other (specify)

Hardware

Are the hardware requirements clearly stated in the course?

❑   Yes

❑   No

Check for all hardware requirements. If appropriate, note the specifications for each component. (Note: In the specifications section of the table, you can add as much information as possible for each component. For example, for the hard disk's *size,* specify how many gigabytes is required or recommended; for the CD ROM drive, specify the required or recommended *speed* 24x or 32x; for RAM, specify the required or recommended *memory size,* 32, 64, 128 or 256 MB; for the monitor, specify the required or recommended *resolution* 640X480, 800X600 or 1024X768, etc.)

Hardware	Check if Required	Check if Recommended	Specifications
Computer and Peripherals			
CPU			
RAM			
ROM			
Hard disk			
Monitor			
Disk drive			
CD-ROM			
CD burner			
Sound card			
Speaker			
Microphone			
Camera			
Video card			
Modem			
DVD (Digital Versatile Disc)			
Ink-jet printer			
Laser printer			
Other (specify)			
Internet Connection			
Dial-in			
DSL (Digital Subscriber Line)			
Cable modem			
T1*			
T3*			
Ethernet			
Wireless connection			
Other (specify)			
Conferencing Tools			
Digital camera			
Video camera			
Other (specify)			
Other Tools			
Cell Phone			
Pager			
PDA (Personal Digital Assistant)			
eBook reader			
Screen reader			
Other (specify)			

* *T1 (DS-1): High-speed digital data channel that is a high-volume carrier of voice and/or data. Often used for compressed video teleconferencing. T-1 has 24 voice channels. T-3 (DS-3): A digital channel that communicates at a significantly faster rate than T-1. A screen reader is a computer software that speaks text on the screen. Often used by individuals who are visually impaired (http://www.learningcircuits.org/glossary.html).*

Does the course require learners to use any new hardware not originally listed in the technology requirement for the course?

❑   Yes
❑   No
❑   Not applicable

If *yes*, are the learners informed?
❑   Yes
❑   No
❑   Not applicable
❑   Other

Check if learners receive training in any of the following. Check all that apply:
❑   How to operate a microphone
❑   How to talk on the microphone
❑   How to do audio conferencing on a PC
❑   How to do video conferencing on a PC
❑   Other (specify)
❑   Not applicable

Does the course provide for desktop videoconferencing or any other type of real time interaction?

❑   Yes
❑   No
❑   Other (specify)
❑   Not applicable

Does the course provide links to resources where learners can learn more about required hardware and their pricing?

❑   Yes
❑   No
❑   Other (specify)
❑   Not applicable

Does the course provide any recommendations on best place(s) to buy various *hardware* components required for the course? (Note: Any such recommendations must be done without any bias or preference. A survey of learners on such issues can be conducted and the results can be posted on the course Website. Also, reviews of hardware from magazines can be useful in this regard. Neither the instructor nor the institution should endorse or promote any particular product. However, if a hardware company is a sponsor or has a special arrangement with the institution to offer special prices for students, then it is a different issue.)

❑ Yes

❑ No

❑ Not applicable

Does the course inform students that the video clips or streaming media[1] (if any) used in the course may not run effectively with a slow modem?

❑ Yes

❑ No

❑ Not applicable

Does the course allow learners to choose any of the following connection speeds for any streaming media used in the course? (Note: With most production software, one can output for different connection speeds. However, if various connection speed options are not provided, Powell (2001) recommends that designers stream media at a low data rate so that individuals with a low connection speed can view it).

❑ 28.8 K

❑ 56K

❑ T1

❑ Not applicable

Software

Are the software requirements for the course clearly stated?

❑ Yes

❑ No

❑ Not applicable

If *yes*, indicate specific software name and check all that apply:

Software	Software Name	Required For			Recommended For			NA
		Learner	*Instructor*	*Other*	*Learner*	*Instructor*	*Other*	
Word Processor								
Email Package								
Presentation Program								
Spreadsheets								
Database								
Graphic Software								
eBook Reader Software								
Audio Video Editing Software								
Operating System								
Plug-ins								
Browsers								
Other (specify)								

Does the course add any new software not originally listed in the technology requirement for the course? (Note: Sometimes, instructor may add a new software after the course is started. This may not be well received by some students as it was not listed before.)

❑   Yes

❑   No

❑   Not applicable

If *yes*, are the learners informed?

❑   Yes

❑   No

❑   Not applicable

❑   Other

Are any browser "plug-ins" needed to use the pages?

❑   Yes

❑   No

❑   Not applicable

If *yes*, are they commonly used and free (such as Acrobat)?

❑   Yes

❑   No

❑   Not applicable

If plug-ins are necessary, is there a link to download them?

❑   Yes

❑   No

❑   Not applicable

❑   Other (specify)

Does the course provide links to resources where learners can learn more about required software and their pricing?

❑   Yes

❑   No

❑   Not applicable

Does the course provide any recommendation on the best place(s) to buy the required *software* for the course? (Note: Any such recommendations must be done without any bias or preference. A survey of learners on such issues can be conducted and the results can be posted on the course Website. Also, reviews of software from magazines can be useful in this regard. Neither the instructor nor the institution should endorse or promote any particular product. However, if a software company is a sponsor or has special arrangement with the institution to offer special prices for students, then it is a different issue.)

❑   Yes

❑   No

❑   Not applicable

Does the course provide links to resources where all necessary software can be downloaded or purchased?

❑   Yes

❑   No

❑   Not applicable

If there are applets or other software to download, does the course specify operating system, memory, CPU and bandwidth requirements?

❑   Yes

❑   No

❑   Not applicable

Check if the course provides any of the following interaction or communication mechanisms for students. (check all that apply):

❑   Chat

❑   E-mail

❑   MUD (Multi-User Dungeon or Dimension)[2]

❑   MOO (Mud, Object Oriented)

❑   Discussion Forum

❑   Newsgroup

❑   Whiteboard

❑   Other (describe below)

Is the course developed using any of the following software? (check all that apply):

❑   LMS

❑   LCMS

❑   Authoring software

❑   Other (specify)

If *yes*, check if the software is in compliance with any of the following standards. (check all that apply):

❑   Institute for Electrical and Electronic Engineers (IEEE)

❑   Instructional Management Systems (IMS)

❑   AICC (Aviation Industry CBT Committee)

❑   SCORM (Sharable Courseware Object Reference Model)

❑   All of the above

❑   Other (describe below)

Are there any criteria used to select LMS, LCMS or the authoring tool? (Note: An article entitled "Selecting a Learning Management System" is available at: http://www.e-learninghub.com/articles/learning_management_system.html. Also, a resources site entitled "Selecting and Using Tools" is available at: http://www.e-learningcentre.co.uk/eclipse/Resources/default-selecting.htm which provides information about LMS, LCMS and authoring tools.).

❑ Yes

❑ No

❑ Not applicable

❑ Other (specify)

If appropriate, check the functionality of the LMS used at the institution. (Note: adopted from http://www.bctechnology.com/statics/pstacey-oct2601.html. Check all that apply:

❑ Schedules and registers learners into online and offline courses

❑ Keeps learner profile data

❑ Launches e-learning courses

❑ Tracks learner progress through courses

❑ Manages classroom based learning

❑ Provides learning administrators with the ability to manage learning resources including labs and classrooms (resource management)

❑ Supports learner collaboration

❑ Automates use of competency maps to define career development and performance paths (skills gap analysis)

❑ Creation of test questions and administration of test

❑ Performance reporting learning results

❑ Interconnectivity with Virtual Classroom (VC), LCMS and enterprise applications

❑ Other (specify)

If appropriate, check the functionality of the LCMS used at the institution. (Note: adopted from http://www.bctechnology.com/statics/pstacey-oct2601.html. Check all that apply:

- ❑ Content migration and management
- ❑ Content creation tools
- ❑ Workflow tools to manage content development process
- ❑ Learning object repository
- ❑ Organizing reusable content
- ❑ Content reuse and adaptive individualized learning paths based on learning objects
- ❑ Asynchronous collaborative learning including discussion groups
- ❑ Testing and certification
- ❑ Reporting of results
- ❑ Delivering content in multiple formats (online, print, PDA, CD-ROM, etc.)
- ❑ Providing content navigational controls (look and feel)
- ❑ Interconnectivity with Virtual Classroom, LMS and enterprise applications

Does the LMS, LCMS or other software used in creating the course work with newer versions of a variety of browsers?

- ❑ Yes
- ❑ No
- ❑ Not applicable
- ❑ Other (specify)

Does the institution use enterprise application software?

- ❑ Yes
- ❑ No
- ❑ Not applicable
- ❑ Other (specify)

If *yes*, can LMS, LCMS or authoring tool used for e-learning be integrated with the institution's enterprise software?

❑ Yes

❑ No

❑ Not applicable

❑ Other (specify

Endnotes

1   Streaming media (streaming audio or video): Audio or video files played as they are being downloaded over the Internet instead of users having to wait for the entire file to download first. Requires a media player program. (Source: http://www.learningcircuits.org/glossary.html#S)

2   http://www.pit.ktu.lt/HP/coper/kiev.new/cit/gloslz.htm#MUD

Chapter 5

Pedagogical Issues

The pedagogical dimension of e-learning encompasses a large set of issues relating to teaching and learning: content analysis, audience analysis, goal analysis, media analysis, design approach, and organization, learning strategies and blending strategies. The following is an outline for the chapter:

- Content analysis
- Audience analysis
- Goal analysis
- Medium analysis
- Design approach
- Instructional strategies
- Organization
- Blending strategies

Content Analysis

Content refers to the subject matter within a domain of knowledge to which a lesson or course is devoted. It also refers to those disciplinary practices that guide the creation, use, and communication of subject-matter knowledge. The content presented within an e-learning unit course depends on the learning goals for that course (see *Goal Analysis* section later in this chapter).

Content analysis involves the identification of the following:

- Content units, sometimes called "chunks," that support learning and facilitate content reuse in other contexts and courses.
- A sequence of content units that serves learners' needs while respecting the integrity of the domain.

Usually, content analysis is carried out by subject matter experts. See *Design Approach* section later in this chapter for the distinction between "ill-structured" and "well-structure" domains.

What types of content are appropriate to teach online? To learn what works and what does not work in e-learning, designers need to take part in the ongoing review of case studies reporting the successes and failures of online learning initiatives across subject areas. Not all content is suited for the electronic environment. As Singh and Reed (2001) state, "Some forms of content — for example, intense behavioral modification, complex physical skills — might only be effectively delivered in face-to-face formats." Rosenberg (2001) helps us understand the appropriateness of contents for e-learning and classroom:

For example, laying a foundation for a house is probably best taught on the job by a skilled craftsman. But if a construction worker someday wants to own his/her own home-building business, she/he better learn a little bit about architecture, accounting, and small business management — clear candidates for e-learning. (p. 125)

Content analysis can thus help designers determine which aspects of a domain's content are appropriate for e-learning and which are appropriate for the face-to-face classroom instruction. Content best served by personal interaction should be taught face-to-face, while content suitable for e-learning should be taught online. Depending on the situation, blended learning (discussed in the *Blended Strategies* section in this chapter) format can be defined as the integration of the best of e-learning offerings with the best of classroom learning offerings. Blended-learning is sometimes referred to as hybrid courses. There-fore, blended learning complements e-learning.

In designing e-learning, we need to consider the stability of course content. Content that does not need to be updated can be categorized as static (e.g., historical events, grammar rules, etc.). Content that has the potential to change over time can be considered dynamic (e.g., laws, policies, etc.). Because dynamic content needs to be revised from time to time, it is necessary to identify such content in a course and establish an ongoing method for timely updating as needed. It will be very frustrating for learners if they find outdated or obsolete information. Therefore, it is important to list all dynamic and stable contents.

Once the suitable content for the e-learning is identified, the next step is to identify content types, including facts, concepts, processes, principles, and procedures. Task analysis must be conducted to identify all tasks and categorize them by type. Content and task analysis are critical in designing learning systems because choices of design approach, content organization, and methods and strategies for learning environment are based on those analyses (Van Merrienboer, 1997; personal communication, April 16, 2000).

Audience Analysis

Who are the e-learners? We can learn about our audience and their specific characteristics by conducting an audience or learner analysis. A careful analysis of learners will provide important information that we can use to design learning activities for our target population. Since e-learning can, in principle, be delivered to anyone, anytime, and anyplace, learners may come from culturally diverse background and they may differ in how they learn. Information about learners' knowledge and skills, personal and social characteristics, capabilities, preferred learning styles, needs, and interests are critical elements of audience analysis (Kemp, Morrison and Ross, 1994). Willis (1992) stated:

To better understand the distance learners and their needs, consider their ages, cultural backgrounds, interests, and educational levels. In addition, assess their familiarity with the various instructional methods and delivery systems being considered, determine how they will apply the knowledge gained in the course, and note whether the class will consist of a broad mix of students or discrete subgroups with different characteristics (e.g., urban/rural, undergraduate/graduate) (http://www.ed.gov/databases/ERIC_Digests/ed351007.html).

People with disabilities merit special attention in this context. Considering the flexibility and convenience of e-learning environment, it is very possible that an increasing numbers of working adults and individuals with disabilities will be considering e-learning. At the Open University of the United Kingdom, the number of students with disabilities is increasing at the rate of 10 percent per year (Thompson, 1998). In designing e-learning for diverse population, it is important to collect and analyze as much information as possible about e-learners, especially working adults and individuals with disabilities.

Various data collection techniques such as surveys, interviews, observations, and document reviews can be used to collect data relating to several learner characteristics, including (but not limited to):

- Age
- Educational level
- Grade-point average
- Standardized test scores
- Cultural background
- Physical and learning disabilities
- Learners' interest
- Experience
- Personal goals and attitudes
- Learning preferences
- Preferred learning styles
- Motivation
- Writing skills
- Reading skills
- Mathematical skills
- Communication skills
- Keyboarding skills
- Word processing skills
- Ability to work with culturally diverse learners
- Familiarity with various instructional methods
- Familiarity with various instructional delivery systems
- Previous experience with e-learning

The more information from these categories is available, the better the e-learning designers will understand their target population. (Please note that institutions may not have adequate information on some issues discussed above such as students' learning preferences, attitude and motivation level.) With comprehensive learner information, designers can design appropriate learning materials for their target audience. The learner analysis can also identify learners for whom the e-learning may not be appropriate.

Goal Analysis

Through the needs assessment process (see *Needs Assessment* in Chapter 2), we can identify what students need to learn, which in turn can help us determine appropriate learning goals. A goal is an expression of instructional purpose, while an objective identifies the performance standards that a student must meet to reach a goal. The goal analysis process helps to identify and clarify the aims of an e-learning project. Identifying clear goals can help learners achieve the greatest learning gains in the most cost-effective and meaningful manner. Clarifying goals can affect the way content is selected and arranged in a specific course (see *Content Analysis* section earlier in this chapter).

In e-learning, it is important for learners to have clear goals and objectives, as well as reasonable ways to achieve them. This section reviews the presence and clarity of those goals and objectives. Smith and Regan (1999) state learning goals are generally more inclusive and less precise than learning objectives. In a course, there can be several units and each unit can have one or more lessons. "We generally do not write a learning goal for a segment smaller than a lesson. However, a lesson may contain many objectives that must be learned to achieve the lesson goal" (Smith & Regan, 1999, p. 65).

Media Analysis

Medium is the means by which the instructional message is communicated. E-learning can be delivered through different media, including the Internet and other digital technologies. In addition, media such as books and printed materials can be combined with e-learning. Media analysis is undertaken to show how media attributes and resources can facilitate learning (Khan, 1997). E-learning designers should be familiar with the capabilities of delivery medium so that they can use them whenever appropriate. Multimedia presentation components such as text, graphics, animation, audio, video, and so on can be used with any e-learning delivery medium to support students in attaining learning goals.

The content analysis process guides the selection of both presentation modes and delivery media for course content. Stable content (e.g., Martin Luther King's "I have a dream" speech) with large audio, video, and graphic files can be delivered to learners via CD-ROM, DVD, or video tape. Or, students could download such content from the course Web site and run it locally from their computers.

Alternative delivery media such as CD-ROM can have a role in Internet-delivered e-learning. Designers should analyze the attributes of each delivery medium (e.g., Internet, video cassette, CD-ROM, DVD, and face-to-face instruction, as well as print-based materials such as books, articles, and manuals) in order to explore whether they can be used for the types of e-learning content that we want. If we want to design a blended e-learning lesson with multiple delivery systems, we have to analyze the capabilities of each medium and then decide which medium is best suited for a specific part of the lesson. For example, an e-learning course in Civil Rights could deliver the "I Have a Dream" speech via either CD-ROM or DVD.

Design Approach

The design approach for e-learning activities is dependent on the type of domain of knowledge of the e-learning content. Jonassen (1997) argues that real world problems are either ill-structured or well-structured. An ill-structured problem, such as figuring out whether or not to trade in your 10-year-old car, losing weight, and so on has multiple solutions or methods for solving it. A well-structured problem, by contrast, such as solving a quadratic equation, can be solved by applying a limited number of rules and principles within well-defined parameters. Jonassen distinguishes between well-structured and ill-structured problems and recommends different design models because they call on different kinds of skills.

The pedagogical philosophy of the overall design of the course is influenced by whether the content is well-structured or ill-structured. The instructivist philosophy espouses an objectivist epistemology, whereas the constructivist approach emphasizes the primacy of the learners' intentions, experiences, and cognitive strategies (Reeves & Reeves, 1997).

In discussing problem solving on the Web, Jonassen, Prevish, Christy, and Stavrulaki (1997) argue that:

... The fundamental difference between constructivist learning environment and objectivist instruction is that here the problem drives the learning. In objectivist instruction, problems function as examples or applications of the concept and principles previously taught. Students learn domain content in order to solve the problem, rather than solving the problem in order to apply the learning. (p. 51)

Control of Learning: An important theoretical consideration for the design of courses on the World Wide Web is the control of learning activities. In designing e-learning activities, we face many design issues such as "should students be in charge of their own learning?" David Peal, a friend and a critical reviewer of my works, thinks that students should not be in charge of all of their own learning when it comes to topics such as fire safety, learning to fly a plane, and so on (personal communication, February 1, 2003). There are two types of control of learning activities: student-centered and program-centered learning activities. As Bannan and Milheim (1997) put it:

Student-centered learning is demonstrated by students selecting and sequencing educational activities as well as creating their own learning opportunities and satisfying their own learning needs (Hooper & Hannafin, 1991). In contrast, program-centered activities involve courses that are highly structured and organized by course designers for the student to later follow, with participation by the student often being prespecified by the instructor and designed to ensure mastery of particular content. Each of these theoretical frameworks significantly impacts the resulting educational methods which are utilized for instructional delivery on the World Wide Web. (p. 382)

Organization

E-learning content should be organized with sequencing strategies (ordering of content) to help learners achieve their goals and objectives.

Instructional Strategies

A variety of instructional strategies can be used in e-learning to facilitate learning and help students achieve their own learning goals and objectives. The strategies used in an e-learning are based in part on the philosophical approach of the course. However, learners' preferences for specific instructional methods are influenced by their learning styles. Using multiple instructional activities can facilitate learning, and the technical and structural attributes of the Internet and digital technologies can be used to support these activities.

In this section, I discuss instructional strategies in terms of their usefulness in e-learning activities[1]. I also look at the Internet and digital technologies that can support these activities.

The instructional approaches and strategies included here are applicable to e-learning. However their use may depend on the type of learning domain (well-defined or ill-defined), the goals and objectives of the course, and the philosophical orientation of the course designers. One can argue that debates may make more sense in social sciences than in chemistry. Leshin, Pollock and Reigeluth (1992) stated,

"Many instructional strategies may work; that is, they may eventually result in the desired learning. Our interest is in selecting optimal strategies—that is, strategies that work better than any others of which we are aware. To select methods, we must have some basis for selection." (p. 3)

In an open, flexible and distributed environment, it is critical to provide learning materials in ways that are accessible to learners with a variety of learning styles.

The e-learning strategies presented in this chapter are not intended to be exhaustive. Instead, my intention is to provide some examples of learning strategies that can be incorporated into e-learning. The following e-learning strategies are discussed in the chapter:

- Presentation
- Exhibits
- Demonstration
- Drill and practice
- Tutorials
- Storytelling
- Games
- Simulations
- Role-playing
- Discussion

- Interaction
- Modeling
- Facilitation
- Collaboration
- Debate
- Field trips
- Apprenticeship
- Case studies
- Generative learning
- Motivation

Presentation

Presentation is defined as a set of techniques for presenting facts, concepts, procedures, and principles. An e-learning presentation can be created using one or more online presentation modes such as text, graphics, photographs, audio clips, video clips, animations, PowerPoint slides, and video-conferencing. Supplemental (offline) materials such as print-based materials, audio, videotapes, CD-ROM, DVD, and so on can be mailed to learners. Online presentations should follow design principles such as keeping things simple, avoiding overcrowding the screen with text and other multimedia components, and ensuring that any presentations made with presentation software run smoothly in different hardware and software configurations.

Exhibits

Exhibits are display objects and visuals for instructional purposes (Heinich, Molenda & Russell, 1993). In e-learning, digital exhibits can be aligned with instruction goals and objectives. Students can use digital exhibits in their projects, which can be an exciting and motivating learning experience for them.

National Gallery of Art Web site at http://www.nga.gov/exhibitions/exhibits.htm hosts virtual art exhibits. The Library of Congress international gallery also hosts virtual exhibits at http://www.loc.gov/exhibits/world/earth.html.

Demonstration

A demonstration is a method of showing or simulating how something works. Demonstrations can be used in e-learning in areas such as teaching procedures, indicating how to operate equipment, illustrating principles, and demonstrating interpersonal skills. Many illustrated demonstrations can be found at the popular Web site, ExploreScience.com (www.explorescience.com).

Drill and Practice

Drill and practice is defined as a learning activity that helps learners master basic skills or memorize facts through repetitive practice. It is most commonly used in teaching math facts, foreign languages, vocabulary (Heinich et al., 1993), reading comprehension, basic science, middle-school history, and geography (Newby, Russel, Stepich & Lehman, 1996). A Web-based drill-and-practice program can provide immediate feedback to learners' responses to various problems presented to them. HTML, Javascript and other scripting languages can be used to create Web-based drill and practice.

Tutorial

A tutorial is a presentation-response-feedback format often used for presenting how-to procedures in the context of a worked example. Web-based tutorials tend to present content, pose questions or problems, ask learners to respond, and finally provide appropriate feedback. For example, free desktop and office tutorials on the Web can be found at: http://www.intelinfo.com/office.html, and a set of programming tutorials can be found at http://www.eng.uc.edu/~jtilley/tutorial.html.

Storytelling

Storytelling is a narrative technique that can be used effectively in e-learning for all cultures. In many cultures, storytelling is used as an educational learning strategy. Stories provide a memorable, compelling format for transferring information and discoveries (Brown, Collins & Duguid, 1989). McLellan (1999) states that stories are a form of "expert system" for remembering and integrating what we learn. Digital storytelling has become a common technique in e-learning. Mellon (1999) reports, "A growing literature on digital storytelling provides a broad definition of the term that incorporates all available multimedia tools — graphics, audio, video, animation, and Web publishing-into the telling of stories" (p. 46). The Center for Digital Storytelling (http://www.storycenter.org/) provides a clearinghouse of information about resources on storytelling and new media.

Games

Games can be a highly motivational instructional device to help learners improve various skills such as decision-making, problem-solving, interpersonal communication, leadership, and teamwork (Newby et al., 1996). In a game, learners follow prescribed rules to attain a challenging and compelling goal. Various Internet and digital technology tools can be used to create games. Thiagarajan and Thiagarajan (2001) use Internet tools such as e-mail, chat-room, and discussion list to create Web-based games. The Play for Performance site (http://thiagi.com/pfp/IE4H/january2003.html) provides useful information about on-line games. Examples of game-based learning approach in e-learning can also be found at Games2Train.com (http://www.games2train.com).

Simulations

Simulations are artificial recreations of real-life situations (Gordon, 1994; Newby et al., 1996). In a simulated environment, learners can practice and make realistic decisions and explore the consequences of their decisions. E-learning can use simulations to improve learners' cognitive, affective, decision-making, and interpersonal skills. Pappo (2001) examines various aspects of design consideration of Web-based simulations.

Examples: The Website Medical Simulations site (http://www.medicalsimulations.com) provides online interactive continuing medical education case studies for physicians and nurses. The Website Rover Ranch: K-12 Experiments in Robotic Software (http://prime.jsc.nasa.gov/ROV/) is an interactive tool that allows students to design and test their own virtual robots.

Role Playing

Role playing can be used to represent real situations that provide learners the opportunity to practice situations they face in the real world (Rothwell & Kazanas, 1992) or to empathize with decision makers, historical figures, and others. Learners can imagine that they are other people in different situations, then make decisions as situations change (Heinich et al., 1993). Role-playing allows learners to learn social skills such as communication and interpersonal skills. In Web-based learning, simulated role portrayal can be facilitated through Multi-User Dialogue (MUD) environments where instructors create a multi-user space with a central theme, characters, and artifacts (Bannan & Milheim, 1997).

The Web site http://www.roleplaysim.org/demos/default.htm provides several examples of role play simulation.

Discussion

Discussion allows learners to analyze information, explore ideas, and share feelings among themselves and their instructors. They can establish communication on the basis of shared interest, not merely shared geography (Harasim, 1993). A well-designed discussion forum in an e-learning course can create an active, interactive, and participatory learning environment. Participants in a discussion forum experience multiple perspectives on issues that encourage them to analyze and appreciate alternative ways of thinking. Therefore, participants in a well-designed discussion forum have the potential to become better critical thinkers.

Online discussions can be either asynchronous (communications are sent and received at different times) or synchronous (communications are sent and received at virtually the same time). Asynchronous text communication tools include e-mail, mailing lists, and newsgroups. Synchronous communication tools include messaging tools and audio- and videoconferencing tools.

In e-learning, learners can be engaged in asynchronous discussions in three different formats: moderated discussion forums, unmoderated discussion forums, and subject-related outside professional discussion forums.

In both synchronous and asynchronous discussions, students not only learn from their instructors, who provide content expertise and feedback during ongoing discussions, but also from each other's comments and feedback. Questions designed to generate and facilitate effective online discussion for instructional purpose should be planned (Berge & Muilenburg, 2000). The online environment designed in a way that promotes open communication while preventing abusive and other non-constructive criticism (Hill & Raven, 2000).

Learners should be reminded of the "code of civility for online discussions." The University of Maryland University College maintains a useful Web site on "Code of Civility and Advisor Confidentiality" (http://www.umuc.edu/studserv/civility.html).

Without body language and eye contact in online text-based discussions, it is very easy for learners to feel isolated and become concerned about what others are thinking (Boehle, 2000). Sometimes, miscommunication in the discussion forums can occur based on how one interprets the meaning of a particular communication. Communication is more open to misinterpretation. It is truism that people from diverse backgrounds will draw different conclusions from the same message. According to Lewis (2000), "Two people can receive the same communication and, as a function of coming from very different backgrounds, can reach two very different conclusions about the meaning of a particular communication" (p. 214).

Instructors/facilitators should be aware of the fact that there are some learners who may take a longer time to post or respond to a message on a discussion forum. Some students who cannot write clearly or quickly may thus be unable to participate actively in online discussion. Non-native speakers in particular may take longer than other students in posting and responding. However, all participants in the discussion forum should develop mutual respect and patience. Layton Montgomery (2000) at the University of Wollongong in Australia shares his observation about non-native speakers' in online asynchronous discussions:

For instance, in English-medium classes, I keep hearing that non-native speakers are less likely to participate in face-to-face discussions, and especially Asian students. It is not that they have less to say, though. The medium is not as conducive to many of them to express their views. When discussion is online and asynchronous, though, these students have the time to consider what is being said more carefully, and respond without feeling embarrassed about their spoken English not being sufficient, and/ or not having to worry about speaking out of turn or inappropriately interrupting somebody else. (eModerators discussion forum, Sub: public dialogue & learning, Dec.11, 2000)

Bailey and Luetkehans (2001) provide the ground rules for a threaded discussion (both Bailey and Luetkehans have experience in facilitating online discussions in their academic courses at Northern Illinois University) in Box 1.

Many discussion forum members suggest that it is always a good idea to either link or add list etiquette and unsubscribe instructions at the end of each message posted. Muilenburg and Berge (2001) in their article, A framework for designing questions for online learning, provide a framework for designing the initial questions for starting online discussions and the follow-up questions for maintaining them. Discussion questions gathered from experienced online instructors are presented with the goal of preparing students and teachers to participate effectively in online discussions (http://www.emoderators.com/moderators/muilenburg.html).

Box 1.

Ground Rules

☑ All ideas are welcome. Honesty and critical reflection are valued.

☑ Participate frequently. Both reading and responding activities are vital to this discussion.

☑ Build on each others' ideas. This is the central strength of a forum. Feel free to question, react to, and build on each others' thoughts.

☑ Let yourself discover a personal style for reading and responding. I encourage you to start by looking for a thread that interests you. Start a new thread if your idea is different from those already posted.

☑ Please use e-mail for private communication.

☑ Please no personal attacks or flaming.

☑ Facilitators will monitor the forum throughout the discussion.

Thiagarajan (2000) recommends, "Much of the feedback on learner assignments can be standardized and reprocessed. Facilitators can post general feedback on a Web page to reduce the time spent in providing individual feedback. Such feedback may include a list of major misconceptions revealed, sample answers with exemplary characteristics, and FAQs" (eModerators discussion forum, Sub: Class size, Dec 3, 2000).

If the learners are not familiar with how a discussion forum works, then a considerable proportion of both learners' and moderators' time will be spent on troubleshooting and technical problems instead of actual discussions of course topics. Therefore, it is necessary for all discussion participants to receive orientation on how to get ready for online discussions.

Interaction

Engagement theory which is based on online learning suggests that students must be meaningfully engaged in learning activities through interaction with others and worthwhile tasks (Kearsley & Shneiderman, 1999). Students can interact with each other, with instructors, and with online resources. Instructors and experts may act as facilitators. They can provide support, feedback, and guidance via synchronous communication (e.g., e-mail and mailing lists) and asynchronous communications tools (e.g., conferencing and messaging tools). Asynchronous communication allows for time-independent interaction whereas synchronous communication tools allows for live interaction (Khan, 1997, 1998).

Depending on the pedagogical philosophy of the course design, both asynchronous and synchronous communication methods can be employed. However, course designers should consider their logistical, instructional, and economic advantages and disadvantages (Hannum, 2001; Berge, Collins & Fitzsimmons, 2001). Instructors should develop skills for promoting online discussion, devising learning activities that work at a distance, and encouraging interaction among the participants (Romiszowski & Chang, 2001). For both asynchronous and synchronous discussions, the instructor or moderator should gently enforce rules and decorum of the discussion forum while encouraging students to engage in vigorous discussions on their own.

Modeling

Modeling is an instructional method through which learners improve their skills by observing and emulating a role model. Modeling provides learners with an example of the desired performance (Jonassen, 1999). It can help learners reach a desired level of performance, deeper understanding, and better grasp of the concept.

Various modeled performances can be used for e-learning activities, ranging from modeling behavior in electronic communication environments to providing samples of relevant coursework. In a Web-based course, posting by instructor of sample interactions, assignments, and projects can provide the necessary modeling for expectations of course requirements (Bannan & Milheim, 1997). "Expert modeling" involves an expert showing how particular problems are solved or how particular situations are handled.

Facilitation

Mentors' and instructors' activities that serve to guide students, direct discussions, suggest possible resources, and field questions (Bannan & Milheim, 1997) are known as facilitation. In e-learning, facilitation can be provided using various tools such as e-mail, mailing lists, discussion forums, and conferencing tools. Frequent and consistent feedback in the online learning can stimulate active engagement by techniques such as questioning assumptions, disagreeing with certain points, and pointing out well-analyzed points (Bischoff, 2000). In the online discussion, the facilitator can ask students questions, suggest alternative perspectives to consider and extend their ideas.

We should not confuse "facilitator" with "moderator." A moderator follows the guidelines established by the institution (see *Etiquette* and *Legal Issues* sections in Chapter 6). The following excerpts posted by DiannaMB on the eModerator mailing list distinguish moderators and facilitators:

My background has been in building and maintaining communities. As such, I have acted as a facilitator, not a moderator. The difference, in my experience, is this: facilitators nurture and grow a community, while moderators police it. To take this one step further — facilitators work for the community (as a servant-leader) and let the community members themselves take ownership of the community under established guidelines (even trusting the community to help establish those guidelines). Moderators, on the other hand, take ownership of the community and act as the overseer — establishing guidelines and expecting them to be followed in a very structured way (eModerators@egroups.com, Sat, 30 Dec 2000).

Williams, Watkins, Daley, Courtenay, Davis, and Dymock (2001) conducted a research study with five faculty members who were part of a cohort research group and who facilitated in an online cross-cultural (cross-international and cross-cultural diversity) environment. Their research noted that, when the instructors facilitated in a cross-cultural online environment, challenges were intensified and expanded beyond the general issues of cultural context.

Collaboration

Collaboration allows learners to work and learn together to accomplish a common learning goal. In a collaborative environment, learners can develop social, communication, critical thinking, leadership, negotiation, interpersonal, and cooperative skills by experiencing the perspectives of other group members. The Web offers extensive opportunities for collaborative learning (Harasim, 1990).

Two types of collaboration can be implemented on the Internet: inside collaboration and outside collaboration. Inside collaboration provides a supportive environment for asking questions, clarifying directions, suggesting or contributing resources, and working on joint projects with class members. Outside collaboration provides for the integration of external personnel and resources, such as speakers, guest lecturers, and Web sites, in course activities (Bannan & Milheim, 1997). E-mail, discussion forums, and conferencing tools can be used to facilitate either kind of collaboration.

Debate

Debate can be used in e-learning to create an authentic learning environment. Debates on controversial issues can help learners engage in a meaningful learning experience. Debate requires that learners select a position and develop an argument to defend it. Debate topics should be based on issues closely related to course content. Debates in e-learning should be designed to promote an open, tolerant, honest exchange of ideas. In debates, arguments should be carried out in accordance with the agreed-upon rules set by the course. Graphics, photographs, audio, video, and discussions can be integrated into debates. Learners can write their own opinions and learn about others'.

A Web site titled "Cultural Debates" (http://www.teachtsp2.com/users/temp/cdonline/index.html) allows students to discover connections and differences between a rainforest society, their own culture, and other communities of students.

Field Trips

Field trips are activities that allow learners to explore places or things to which they would otherwise not have access (Khan, 1997). Field trips through the Web allow the instructor to provide students with a guided tour to a city, park, or business Web site as if the instructor were taking students on a field trip (Badger, 2000). In e-learning, students should be provided with themes and objectives for the field trips so that they can gather appropriate information as part of their assigned tasks.

Tramline Virtual Field Trips (http://www.field-trips.org) site has created a range of field trips on nature topics. These trips are particularly well suited to classroom use and provide teacher's objectives and resources for each trip.

Apprenticeship

Apprenticeship offers learners the chance to observe, model, and interact with mentors or experts for particular learning tasks. The conferencing and collaboration technologies of the Web bring students into contact with authentic learning and apprenticing situations (Bonk & Reynolds, 1997). The apprenticeship method in e-learning can help create an ongoing dialogue between mentors and learners that in turn can help learners gain greater knowledge and skill in the area of their interests.

GLOBE, for example, a worldwide education and science program for primary and secondary schools, makes use of the apprenticeship approach (http://www.globe.gov/) using various Internet tools and digital technologies.

Case Studies

Case studies are real or hypothetical situations developed in depth for use in an e-learning course in order to engage learners in realistic problem-solving tasks. Cases can encourage discussion about best practices and problem-solving strategies, and can be based on the actual situations that learners are likely to encounter when they become practitioners (Brown, et al., 1989). These cases should, of course, be aligned with the learning goal(s) of the course in order for learners to benefit from them.

Links to various case study sites on the Internet can be found at Case Studies in Science (http://ublib.buffalo.edu/libraries/projects/cases/webcase.htm).

Generative Learning

Originally designed to improve reading comprehension, the generative-learning model suggests that students achieve comprehension of new material in two phases. First, they create relationships within new material. Second, they build connections between the new information and their existing knowledge, restructuring their knowledge in the process.

"Generative" refers to the generation of new understanding, and in the generative-learning model, learning is always active. Generative-learning is most effective when students have intrinsic motivation, use metacognitive (self-regulating) skills to monitor their progress, and attribute learning to their own effort (David Peal, Personal Communication, February 4, 2003).

The model informs a large set of practical techniques for comprehending and integrating new information. To comprehend new information, students can identify text features (title, headings), make predictions, ask questions, draw diagrams, write summaries, and elaborate upon what they read. To aid the integration of this new information into the memory structures that make up their prior knowledge, students can think of examples, make inferences, devise applications (uses of the new material), demonstrate their new knowledge, and create metaphors and analogies that capture the gist of the new material and relate it to prior, or common, knowledge.

Generative learning requires opportunities for students to use new information actively. These opportunities can take the form of a teacher's prompts for a student to generate connections within new information and between new information and prior knowledge, or they can take the form of questions and activities developed as part of Web-based lessons. Designing for generative learning requires that students be given, for example, the opportunity to ask themselves questions before, during, and after study; to summarize and elaborate upon what they read; and represent the new knowledge through metaphors, analogies, and diagrams.

For a review of the extensive literature on generative learning, see Grabowski (1996). For Web resources, see Martin Ryder's collection of helpful links at http://carbon.cudenver.edu/~mryder/itc_data/idmodels.html#generative.

Motivation

E-learning courses should be designed to motivate students so that they can enjoy their learning experience on the Web and complete their assignments on time. Motivation can be encouraged within any instructional method discussed above. Cornell and Martin (1997) advise course designers to provide a variety of stimuli, varied strategies, and diverse sources of media formats. The e-learning environment should create a positive first impression, be readable, use graphics and pictures that are relevant and useful, provide cues to the learners, and stimulate early interest so that students will be more likely to complete the course. In online asynchronous discussions, an instructor's timely feedback assures learners that they are focusing on their learning, which in turn serves as motivational and beneficial factors to their learning processes (Bischoff, 2000). Also, when the course does not require students to participate on frequent and scheduled online discussions and other online activities (e.g., quizzes), they may procrastinate, which in turn may affect their motivation.

In a pair of national surveys on the state of online leaning, Bonk (2001, 2002) found that most courses are pedagogically void. Fairly high attrition rates are due to the lack of motivating and engaging materials. There is thus a dire need to create online materials and courses that engage learners in interactive and meaningful learning activities instead of merely turning the electronic pages (Dennen, 2001). Based on a series of research studies, practical experience, and a review of the literature on motivation, Dennen and Bonk (in press) identified ten key elements for motivating online learners:

1.   Tone/climate, which is set at the beginning of the course and should engage students and explain expectations as well as enable students to share ideas and personal information.

2. Feedback, which helps students know if they are meeting course expectations and to relate alternative points of view.

3. Engagement, which involves making learners active participants and contributors and excites them into the online environment.

4. Meaningfulness, which can be achieved through use of real-world examples and making connections between course material and students lives.

5. Choice, which involves providing learners with options or alternatives and a sense of control over the learning environment, such as conference tracks for discussions.

6. Variety, which typically is related to providing different learning activities to keep learners interested and attentive.

7. Curiosity, which should be cultivated through elements of surprise, novelty, and intrigue and is encouraged by course extensions and outside perspectives.

8. Tension, a positive term here, which can encourage debate, dissonance, conflict, and the sharing of multiple perspectives; in effect, there is a sense of not knowing something or having a difference of opinion.

9. Peer interaction, which is a method to encourage students to exchange ideas, provide feedback, participate in the course, and review each other's work through which students frequently judge their own progress and come to feel a part of a learning community.

10. Goal-driven, which refers to a student's motivation to participate in a course in order to complete a task, activity, product, or problem and thus feel a sense of accomplishment and personal pride; as a result, the learning activities should be clearly aligned to the course goals.

The discussion of various e-learning methods and strategies in this chapter can help us select appropriate strategies for various parts of an e-learning course. It is always important to identify appropriate methods to enhance learning. We may find several methods appropriate for a specific e-learning content, but we should select the method that best serves our target audience (i.e., learners) within our technological and financial capabilities.

Blending Strategies

While learning technologies and delivery media continue to evolve and progress, we should take advantage of all possible delivery media, whenever appropriate, to design blended learning environments. Many organizations favor blended learning models over single delivery mode programs. Singh (2003) states that a single mode of instructional delivery may not provide sufficient choices, engagement, social contact, relevance, and context needed to facilitate successful learning and performance.

Is there a way to appropriately blend various forms of learning? Singh (2003) notes that the e-learning framework can enable one to select appropriate ingredients for blended-learning, and it can also guide to plan, develop, deliver, manage, and evaluate blended-learning programs. In this section, with the permission from *Educational Technology* magazine, I excerpted the following article titled "Building Effective Blended Learning Programs" by Harvey Singh (2003) to discuss blending strategies.

Building Effective Blended Learning Programs

The first generation of e-learning or Web-based learning programs focused on presenting physical classroom-based instructional content over the Internet. Furthermore, first-generation e-learning (digitally delivered learning) programs tended to be a repetition or compilation of online versions of classroom-based courses. The experience gained from the first-generation of e-learning, often riddled with long sequences of 'page-turner' content and point-and-click quizzes, is giving rise to the realization that a single mode of instructional delivery may not provide sufficient choices, engagement, social contact, relevance, and context needed to facilitate successful learning and performance.

In the second wave of e-learning, increasing numbers of learning designers are experimenting with blended learning models that combine various delivery modes. Anecdotal evidence indicates that blended learning not only offers more choices but also is more effective.

This section has two objectives:

1. To provide a comprehensive view of blended learning and discuss possible dimensions and ingredients (learning delivery methods) of blended learning programs.

2. To provide a model to create the appropriate blend by ensuring that each ingredient, individually and collectively, adds to a meaningful learning experience.

Badrul Khan's e-learning framework, referred to here as the *Octagonal Framework* (see Figure 1) enables one to select appropriate ingredients. The framework serves as a guide to plan, develop, deliver, manage, and evaluate blended learning programs. Organizations exploring strategies for effective learning and performance have to consider a variety of issues to ensure effective delivery of learning and thus a high return on investment.

Blended Learning

Learning requirements and preferences of each learner tend to be different. Organizations must use a blend of learning approaches in their strategies to get the right content in the right format to the right people at the right time. Blended learning combines multiple delivery media that are designed to complement each other and promote learning and application-learned behavior.

Blended learning programs may include several forms of learning tools, such as real-time virtual/ collaboration software, self-paced Web-based courses, electronic performance support systems (EPSS) embedded within the job-task environment, and knowledge management systems. Blended learning mixes various event-based activities, including face-to-face classrooms, live e-learning, and self-paced learning. This often is a mix of traditional instructor-led training, synchronous online conferencing or training, asynchronous self-paced study, and structured on-the-job training from an experienced worker or mentor.

Dimensions of the Blend

The original use of the phrase "blended learning" was often associated with simply linking traditional classroom training to e-learning activities, such as asynchronous work (typically accessed by learners outside the class at their own time and pace). However, the term has evolved to encompass a much richer set of learning strategies or "dimensions." Today a blended learning program may combine one or more of the following dimensions, although many of these have over-lapping attributes.

Blending Offline and Online Learning

At the simplest level, a blended learning experience combines off-line and online forms of learning where the online learning usually means "over the Internet or Intranet" and off-line learning happens in a more traditional classroom setting. We assume that even the off-line learning offerings are managed through an online learning system. An example of this type of blending may include a learning program that provides study materials and research resources over the Web, while providing instructor-led, classroom training sessions as the main medium of instruction.

Blending Self-Paced and Live, Collaborative Learning

Self-paced learning implies solitary, on-demand learning at a pace that is managed or controlled by the learner. Collaborative learning, on the other hand, implies a more dynamic communication among many learners that brings about knowledge sharing. The blending of self-paced and collaborative learning may include review of important literature on a regulatory change or new product followed by a moderated, live, online, peer-to-peer discussion of the material's application to the learner's job and customers.

Blending Structured and Unstructured Learning

Not all forms of learning imply a premeditated, structured, or formal learning program with organized content in specific sequence like chapters in a textbook. In fact, most learning in the workplace occurs in an unstructured form via meetings, hallway conversations, or e-mail. A blended program design may look to actively capture conversations and documents from unstructured learning events into knowledge repositories available on-demand, supporting the way knowledge-workers collaborate and work.

Blending Custom Content with Off-the-Shelf Content

Off-the-shelf content is by definition generic — unaware of an organization's unique context and requirements. However, generic content is much less expensive to buy and frequently has higher production values than custom content. Generic self-paced content can be customized today with a blend of live experiences (classroom or online) or with content customization. Industry standards such as SCORM (Shareable Content Object Reference Model) open the door to increasingly flexible blending of off-the-shelf and custom content, improving the user experience while minimizing cost.

Blending Learning, Practice, and Performance Support

Perhaps the finest form of blended learning is to supplement learning (organized prior to beginning a new job-task) with practice (using job-task or business process simulation models) and just-in-time performance support tools that facilitate the appropriate execution of job-tasks. Cutting-edge productivity tools provide 'workspace' environments that package together the computer based work, collaboration, and performance support tools.

Why Blend?

The Benefits of Blending

Blended learning is not new. However, in the past, blended learning was comprised of physical classroom formats, such as lectures, labs, books, or handouts. Today, organizations have a myriad of learning approaches and choices. Some of these are shown in Table 1.

The concept of blended learning is rooted in the idea that learning is not just a one-time event — learning is a continuous process. Blending provides various benefits over using any single learning delivery medium alone.

However, organizations are not limited to the aforementioned approaches and delivery technologies. Blended learning provides for flexibility that organizations can utilize to their advantage and develop learning programs that suit their requirements better.

Table 1. Learning approaches and choices

Synchronous physical formats	Instructor-led classrooms and lectures Hands-on labs and workshops Field trips
Synchronous online formats (live e-learning)	e-Meetings Virtual classrooms Web seminars and broadcasts Coaching Instant messaging Conference calls
Self-paced, asynchronous formats	Documents and Web Pages Web/computer based training modules Assessments/tests and surveys Simulations Job aids and Electronic Performance Support Systems (EPSS) Recorded live events Online learning communities and discussion forums Distributed and mobile learning

Extending the Reach

A single delivery mode inevitably limits the reach of a learning program or critical knowledge transfer in some form or fashion. For example, a physical classroom-training program limits the access to only those who can participate at a fixed time and location, whereas a virtual classroom event is inclusive of remote audiences and, when followed up with recorded knowledge objects (ability to playback a recorded live event), can extend the reach to those who could not attend at a specific time.

Optimizing Development Cost and Time

Combining different delivery modes has the potential to balance out and optimize the learning program development and deployment costs and time. A totally online, self-paced, media-rich, Web-based training content may be too expensive to produce (requiring multiple resources and skills), but combining virtual collaborative and coaching sessions with simpler self-paced materials, such as generic off-the-shelf WBT, documents, case studies, recorded e-learning events, text assignments, and PowerPoint presentations (requiring quicker turn-around time and lower skill to produce) may be just as effective or even more effective.

Evidence that Blending Works

We are so early into the evolution of blended learning that little formal research exists on how to construct the most effective blended program designs. However, research from institutions such as Stanford University and the University of Tennessee have given us valuable insight into some of the mechanisms by which blended learning is better than both traditional methods and individual forms of e-learning technology alone. This research gives us confidence that blending not only offers us the ability to be more efficient in delivering learning, but more effective.

Stanford University has over 10 years of experience with self-paced enrichment programs for gifted youth. Their problem was that only slightly more than half of their highly motivated students would complete the programs. They diagnosed the problem as a mismatch between the student's desired learning style — interactive, social, mentored learning — with the delivery technology. Their introduction of live e-learning into their program raised the completion rate up to 94 percent by addressing these needs. The improvement was attributed to the ability of a scheduled live event to motivate learners to complete self-paced materials on time; the availability of interaction with instructors and peers; and higher quality mentoring experiences. The Stanford research strongly suggests that linking self-paced material to live e-learning delivery could have a profound effect on overall usage and completion rates — enabling organizations to radically increase the return from their existing investments in self-paced content.

Research by the University of Tennessee's Physician's Executive MBA (PEMBA) program[2] for mid-career doctors has demonstrated that blended learning programs can be completed in approximately one-half the time, at less than half the cost, using a rich mix of live e-learning, self-paced instruction, and physical classroom delivery. Of even greater interest, this well-designed program was also able to demonstrate an overall 10 percent better learning outcome than the traditional classroom learning format — the first formal study to show significant improvements from e-learning rather than just equivalent outcomes. This exceptional outcome was attributed by PEMBA to the richness of the blended experience that included multiple forms of physical and virtual live e-learning, combined with the ability of the students to test their learning in the work context immediately and to collaborate with peers in adaptation to their unique environments.

Taken together, these studies show us that, regardless of your starting point the traditional classroom or self-paced e-learning, the diversity of a blending learning experience appears to have a significant impact on the overall effectiveness of a learning program relative to any individual learning delivery method alone. How do you bring some of these benefits to your organization? First of all, you should not be afraid to get started. We still have years of innovation ahead of us as we explore new learning technologies and program combinations that will generate even better results.

The Octagonal Framework

As discussed in Chapter 1, each dimension in the framework (Figure 1) represents a category of issues that need to be addressed to create a meaningful learning environment. These issues help organize thinking and ensure that the resulting learning program creates a meaningful learning experience.

Institutional

Personnel involved in the planning of a learning program could ask questions related to the preparedness of the organization, availability of content and

Figure 1. The octagonal framework

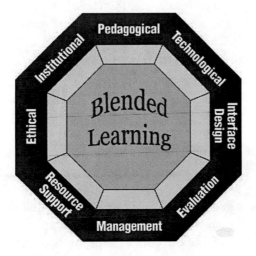

infrastructure, and learners' needs. Can the organization manage offering each trainee the learning delivery mode independently as well as in a blended program? Has the needs analysis been performed in order to understand all learners' needs?

Pedagogical

The Pedagogical dimension is concerned with the combination of content that has to be delivered (content analysis), the learner needs (audience analysis), and learning objectives (goal analysis). The pedagogical dimension also encompasses the design and strategy aspect of e-learning.

This dimension addresses a scenario where all learning goals in a given program are listed and then the most appropriate delivery method is chosen. For example, if a learner is expected to demonstrate a product (in sales training), then using product simulation as part of the blend is appropriate. If a learner is expected to come up with a new price model for a product, then using a discussion as one of the elements in the blend would be an appropriate choice.

Technological

Once we have identified the delivery methods that are going to be a part of the blend, the technology issues need to be addressed. Issues include creating a learning environment and the tools to deliver the learning program. This dimension addresses the need for the most suitable learning management system (LMS) that would manage multiple delivery types and a learning content management system (LCMS) that catalogs the actual content (online content modules) for the learning program.

Technical requirements, such as the server that supports the learning program, access to the server, bandwidth and accessibility, security, and other hardware, software, and infrastructure issues are addressed.

Interface Design

The interface design dimension addresses factors related to the user interface of each element in the blended learning program. One needs to ensure that the user interface supports all the elements of the blend. The interface has to be sophisticated enough to integrate the different elements of the blend. This will enable the learner to use each delivery type and switch between the different types. The usability of the user interface will need to be analyzed. Issues like content structure, navigation, graphics, and help also can be addressed in this dimension.

For example, in a higher education course, students may study online and then attend a lecture with the professor. The blended learning course should allow students to assimilate both the online learning and the lecture equally well. They must be able to distinguish the breaks in the syllabus.

Evaluation

The evaluation dimension is concerned with the usability of a blended learning program. The program should have the capability to evaluate how effective a learning program has been as well as evaluating the performance of each learner. In a blended learning program, the appropriate evaluation method should be used for each delivery type.

For example, to evaluate the participation of learners in a discussion, you can make a qualitative assessment of a learner's contribution to the discussion and understanding of the topic of discussion.

Management

The management dimension deals with issues related to the management of a blended learning program, such as infrastructure and logistics to manage multiple delivery types. Delivering a blended learning program is more work than delivering the entire course in one delivery type. The management dimension also addresses issues like registration and notification, and scheduling of the different elements of the blend.

A typical scenario could be that the blended learning program is available to learners in different time zones (the self-paced online learning part of the program); you need to ensure that the servers are always accessible.

Resource Support

The resource support dimension deals with a blended learning program making different types of resources (offline and online) available for learners as well as organizing them. Resource support could also be a counselor/ tutor always available in person, via e-mail or on a chat. For example, books and periodicals need to be organized in the support center or library (phsical or virtual) so students can find them easily. The URL (Web sites) pointers should be made accessible so students can easily reference the Web sites.

Ethical

The ethical dimension identifies the ethical issues that need to be addressed when developing a blended learning program. Issues such as equal opportunity, cultural diversity, and nationality should be addressed. Learning programs should be designed in way that does not offend any participant.

Blended learning programs should be developed to enable all learners to have a similar learning experience for each of the ingredient of the blended learning program. Alternate choices need to be provided for learner with special needs. The different elements of the blend should also be compatible with each other.

Applying the Octagonal Framework to Blended Learning

Compared with conventional or single-mode e-learning, blended learning programs require more thought and planning. The dimensions of the Octagonal Framework allow us to go through the rigorous planning and design before the blended learning programs are offered to the participants. The framework and its dimensions are relevant for both academic and corporate learning programs. Each dimension of the Octagonal Framework has several sub-dimensions, each consisting of issues focused on a specific aspect of a learning strategy that can further assist us in selecting the right ingredients for the blended learning programs.

Conclusion

While learning technologies and delivery media continue to evolve and progress, one thing is certain: organizations (corporate, government, and academic) favor blended learning models over single delivery mode programs. Blended learning programs must match organizational, learning, and performance objectives with the right ingredients or delivery technologies to achieve optimum results in terms of cost, time, and quality parameters. The Octagonal Framework serves as a tool to help select each delivery option individually and wholistically to create the blended learning programs.

Question to Consider

Can you think of any e-learning related pedagogical issues not covered in this chapter?

Activity

1.   Using Internet search engines, locate and describe at least one example of any of the following instructional strategies used in online courses:

     •   Presentation

     •   Exhibits

     •   Demonstration

     •   Drill and practice

     •   Tutorials

     •   Storytelling

- Games
- Simulations
- Role-playing
- Discussion
- Interaction
- Modeling
- Facilitation
- Collaboration
- Debate
- Field trips
- Apprenticeship
- Case studies
- Generative learning
- Motivation

2. Locate an online program and review its pedagogical aspects using the relevant pedagogical checklist items at the end of Chapter 5.

References

Badger, A. (2000). Keeping it fun and relevant: Using active online learning. In K.W. White & B.H. Weight (Eds.), *The online teaching guide* (pp. 124-141). Boston: Allyn & Bacon.

Bailey, M. & Luetkehans, L. (2001). In B.H. Khan (Ed.), *Web-based training,* (pp. 235-244). Englewood Cliffs, NJ: Educational Technology Publications.

Bannan, B. & Milheim, W.D. (1997). Existing Web-based courses and their design. In B.H. Khan (Ed.), *Web-based instruction,* (pp. 381-388). Englewood Cliffs, NJ: Educational Technology Publications.

Berge, Z.L., Collins, M. & Fitzsimmons, T. (2000). Advantages and disadvantages of Web-based training. In B.H. Khan (Ed.), *Web-based training* (pp. 21-26). Englewood Cliffs, NJ: Educational Technology Publications.

Berge, Z.L. & Muilenburg, L. (2000). Designing Discussion Questions for Online, Adult learning. *Educational Technology*, *40*(5), 53-56.

Boehle, S. (2000). My exasperating life as an online learner. *Training Magazine*, 64-68.

Bonk, C.J. (2001). Online teaching in an online world. Bloomington, IN: CourseShare.com. Retrieved January 1, 2003, from *http://Publication Share.com*

Bonk, C.J. (2002). Online training in an online world. Bloomington, IN: CourseShare.com. Retrieved January 1, 2003, from *http://Publication Share.com*

Bonk, C.J. & Reynolds, T.H. (1997). Leraner-centered Web instruction for higher-order thinking, teamwork and apprenticeship. In B.H. Khan (Ed.), *Web-based instruction,* (pp. 167-178). Englewood Cliffs, NJ: Educational Technology Publications.

Brown, J.S., Collins, A. & Duguid, P. (1989). Situated cognition and the culture of learning. *Educational Researcher, 18*(1), 32-42.

Cornell, R. & Martin, B.L. (1997). The role of Motivation in Web-based instruction. In B.H. Khan (Ed.), *Web-based instruction,* (pp. 93-100). Englewood Cliffs, NJ: Educational Technology Publications.

Dennen, V.P. (2001). *The design and facilitation of asynchronous discussion activities in Web-based courses.* Unpublished doctoral dissertation, University of Indiana, Bloomington.

Dennen, V. & Bonk, C.J. (in press). We'll leave a light on for you: Keeping learners motivated in online courses. To appear in B.H. Khan (Ed.), *Flexible learning.* Englewood Cliffs: Educational Technology Publications.

Gordon, S.E. (1994). *Systematic training program design: Maximizing effectiveness and minimizing liability.* Englewood Cliffs, NJ: Prentice Hall.

Grabowski, B.L. (1996). Generative learning: Past, present, and future. In D.H. Jonassen (Ed.), *Handbook of research for educational communications and technology* (pp. 897-918). New York: MacMillan.

Harasim, L. (1990). Online education: An environment for collaboration and intellectual amplification. In L. Harasim (Ed.), *Online education: Perspectives on a new environment.* New York: Praeger Publishers.

Harasim, L. (1993). (Ed.). *Global networks: Computers and international communication.* Cambridge, MA: MIT Press.

Hannum, W. (2001). Web-based training: Advantages and limitations. In B.H. Khan (Ed.), *Web-based training,* (pp. 13-20). Englewood Cliffs, NJ: Educational Technology Publications.

Heinich, R., Molenda, M. & Russell, J. (1993). *Instructional media and technologies for learning.* Upper Saddle River, NJ: Prentice-Hall, Inc.

Hooper, S. & Hannafin, M.J. (1991). *The effects of group composition on achievement, interaction and learning efficiency during computer-*

based cooperative instruction. Educational Technology Research and Development.

Hill, J.R. & Raven, A. (2000). Online learning communities: If you build them, will they stay? *http://it.coe.uga.edu/itforum/paper46/paper46.htm*

Jonassen, D.H. (1997). Instructional design models for well-structured and ill-structured problem solving. *Educational Technology Research and Development, 45*(1), 25.

Jonassen, D.H. (1999). Designing constructivist learning environments. In C.M. Reigeluth (Ed.), *Instructional theories and models (volume II)*. Mahwah, NJ: Lawrence Erlbaum.

Jonassen, D.H., Prevish, T., Christy, D. & Stavrulaki, E. (1997). Learning to solve problems on the Web: Aggregate planning in a business management course. *Distance Education, 20*(1), 49-62.

Kearsley, G. & Shneiderman, B. (1999). Engagement theory: A framework for technology-based teaching and learning. (*http://home.sprynet.com/ ~gkearsley/engage.htm*).

Kemp, J.E., Morrison, G.R. & Ross, S.M. (1994). *Designing effective instruction*. New York: Merrill.

Khan, B.H. (1997). Web-based instruction: What is it and why is it? In B. H. Khan (Ed.), *Web-based instruction,* (pp. 5-18). Englewood Cliffs, NJ: Educational Technology Publications.

Khan, B.H. (1998). Web-based instruction: An introduction. *Educational Media International, 35*(2), 63-71.

Leshin, C.B., Pollock, J. & Reigeluth, C.M. (1992). *Instructional design strategies and tactics*. Englewood Cliffs, NJ: Educational Technology Publications.

Lewis, C. (2000). Taming the lions and tigers and bears. In K.W. White & B.H. Weight (Eds.), *The online teaching guide*. Boston: Allyn & Bacon.

McLellan, H. (1999). Online education as interactive experience: Some guiding models. *Educational Technology, 39*(5), 36-42.

Mellon, C.A. (1999). Digital storytelling: Effective learning through the Internet. *Educational Technology, 39*(2), 46-50.

Muilenburg, L. & Berge, Z.L. (2001). A framework for designing questions for online learning. *http://www.emoderators.com/moderators/muilenburg. html*

Newby, T.J., Russel, J.D., Stepich, D.A. & Lehman, J.D. (1996). *Instructional technology for teaching and learning: Designing instruction, integrating computers, and using media*. Hillsdale, NJ: Prentice Hall.

Pappo, H.A. (2001). Simulations for Web-based training. In B.H. Khan (Ed.), *Web-based training,* (pp. 225-228). Englewood Cliffs, NJ: Educational Technology Publications.

Reeves, T.C. & Reeves, P.M. (1997). Effective dimensions of interactive learning on the World Wide Web. In B.H. Khan (Ed.), *Web-based instruction,* (pp. 59-66). Englewood Cliffs, NJ: Educational Technology Publications.

Romiszowski, A.J. & Chang, E. (2001). A practical model for conversational Web-based training: A response from the past to the needs of the future. In B.H. Khan (Ed.), *Web-based training,* (pp. 107-128). Englewood Cliffs, NJ: Educational Technology Publications.

Rosenberg, M.J. (2001). *E-learning: Strategies for delivering knowledge in the digital age.* New York: McGraw-Hill.

Rothwell, W. & Kazanas, H.C. (1992). *Mastering the instructional design process.* San Francisco: Jossey-Bass Publishers.

Singh, H. (2003). Building effective blended learning programs. *Educational Technology, 44*(1), 5-27.

Singh, H. & Reed, C. (2001). Achieving success with blended learning. Retrieved January 29, 2003, from *http://www.astdet.org/Centra%20-%20Blended%20Learning.pdf*

Smith, P.L. & Ragan, T.J. (1999). *Instructional design.* New York: John Wiley & Sons.

Thiagarajan, S. & Thiagarajan, R. (2001). Playing interactive training games on the Web. In B.H. Khan (Ed.), *Web-based training,* (pp. 219-224). Englewood Cliffs, NJ: Educational Technology Publications.

Thompson, M.M. (1998). Distance learners in higher education. In C.C. Gibson (Ed.), *Distance learners in higher education.* Madison, Wisconsin: Atwood Publishing.

Williams, S.W., Watkins, K., Daley, B., Courtenay, B., Davis, M. & Dymock, D. (2001). International distance education: Cultural and ethical issues. *Distance Education: An International Journal, 22*(1), 151-167.

Willis, B. (1992). Instructional development for distance education. *ERIC Digest.* ED351007. ERIC Clearinghouse on Information Resources Syracuse NY.

Endnotes

¹   Please note that I maintain a resource Web site titled "E-Learning Methods and Strategies" at http://BooksToRead.com/elearning/strategies.htm which provides links to relevant Web sites dealing with various strategies included in this section.

²   Effectiveness of combined delivery modalities for distance learning and resident learning; P. Dean, M. Stahl, D. Sylwester, & J. Peat; *Quarterly Review of Distance Education*, July/August 2001.

Pedagogical Checklist

Content Analysis

Is the content of the course accurate?
- ❏   Yes
- ❏   No
- ❏   Not applicable
- ❏   Other (specify)

Given the course goals, is the content complete?
- ❏   Yes
- ❏   No
- ❏   Not applicable

If *yes*, list the topics within a content area that would be suitable for e-learning, face-to-face instruction and/or other.

Topics	Content Suitability Analysis		
	E-Learning	Face-to-Face	Other (specify)
1.			
2.			
3.			
4.			
5.			
5.			
6.			
7.			
8.			

List stable and dynamic contents for each lesson. (Note: Content that does not need to be updated can be categorized as *static*. For example, historical events, grammar rules, etc. Content that has the potential to change over time can be considered *dynamic*. For example, laws, policies, etc.).

Lesson Name	E-Learning Content Stability Analysis	
	List *Stable* Contents	List *Dynamic* Contents

How often is the dynamic course content updated?
- ❑ Weekly
- ❑ Monthly
- ❑ Quarterly
- ❑ Yearly
- ❑ As needed
- ❑ Not applicable
- ❑ Other (specify)

Check all that applies for content types in each lesson or unit of the course.

Content Types	Lesson or Unit									
	1	*2*	*3*	*4*	*5*	*6*	*7*	*8*	*9*	*10*
Facts										
Concepts										
Processes										
Principles										
Procedures										
Other (specify)										

Check all that apply for the types of reading assignments recommended for online and offline activities of the course:

Types of Reading Assignments	Online Activity	Offline Activity	Other (specify)
Required textbook(s)			
Readings from books other than textbook(s)			
Readings from printed-based journal/magazine			
Readings from online magazine and journal			
Readings from e-books			
Case study			
Readings located at the course Website			
Readings located in non-course Website			
Readings from CD-ROM			
Reference materials			
Other (specify)			

Does the course require students to do any offline activities (e.g., physical and hands on activities- such as viewing particular TV programs, visiting learning centers, library, etc.)?

❑   Yes

❑   No

❑   Not applicable

If *yes*, please specify:

Does a course's description in the syllabus communicate the importance and relevance of its content?

❑   Yes

❑   No

❑   Not applicable

❑   Other (specify)

Is the description of the course in the syllabus the same as the one approved by the curriculum committee? (Note: E-learning designers should always refer to the approved documents. If any changes are made in the syllabus, learners should be notified.)

❑   Yes

❑   No

❑   Not applicable

❑   Other (specify)

Does the course content use any textual and multimedia materials from outside sources?

❑   Yes

❑   No

❑   Not applicable

❑   Other (specify)

> If *yes*, check all that apply:
>
> ❑   Textual and multimedia materials used in the course represent a variety of viewpoints
>
> ❑   Accurate information about where the materials came from is provided
>
> ❑   Other (specify)

Audience Analysis

Does the institution have any of the following demographic information available about learners? (check all that apply):

Demographic Information	Yes	No	NA	Other
Age range				
Gender				
Educational level				
Grade-point average (GPA)				
Standardized test scores (e.g., SAT, GRE)				
Socioeconomic background				
Racial/ethnic background				
Physical disabilities				
Learning disabilities				

Does the institution have any of the following knowledge and skills information available about learners? (Note: If the institution does not have information about learners' knowledge and skills, e-learning designers should proactively do surveys to gather such information.)

Knowledge and Skills Information	Yes	No	NA	Other
Writing skills				
Reading skills				
Mathematical skills				
Communication skills				
Keyboarding skills				
Word processing skills				
Internet navigation skills				
Previous experience with e-learning				
Ability to work independently				
Ability to work with culturally diverse learners				
Familiarity with various instructional methods				
Familiarity with different delivery systems				
Other (specify)				

If any of the following information is available about learners, check all that apply for their knowledge and skills level:

Knowledge and Skills Information	Level				
	Excellent	Good	Fair	Good	NA
Writing skills					
Reading skills					
Mathematical skills					
Communication skills					
Keyboarding skills					
Word processing skills					
Internet navigation skills					
Previous experience with e-learning					
Ability to work independently					
Ability to work with culturally diverse learner					
Familiarity with various instructional methods					
Familiarity with different delivery systems					
Other (specify)					

Does the institution have any of the following learning preferences information available about learners? (Note: If the institution does not have such information, e-learning designers should proactively do surveys to find out what learners prefer.)

Learning Preferences	Yes	No	NA	Other
Lecture				
Presentation				
Exhibits				
Demonstration				
Drill and Practice				
Tutorials				
Games				
Storytelling				
Simulations				
Role-playing				
Discussion				
Interaction				
Modeling				
Facilitation				
Collaboration				
Debate				
Field Trips				
Apprenticeship				
Case Studies				
Other (specify)				

Does the institution have information about learners' preferred learning styles (e.g., visual/nonverbal, visual/verbal, auditory and kinesthetic)? (Note: If the institution does not have information about learners' preferred learning styles, e-learning designers should proactively do surveys to gather such information. The l2earner diversity section discusses learning styles.)

❑   Yes
❑   No
❑   Not applicable
❑   Other (specify)

If *yes*, write a summary of the target population's preferred learning styles.

Does the institution have any of the following attitudinal and motivational information available about learners? (check all that apply):

Attitudinal and Motivational Information	Yes	No	NA	Other
Motivation level				
Interests				
Anxiety level				
Attitude toward learning				
Attitude toward instructional content of the course				
Learners' expectations concerning the course content				
Learners' expectations concerning instructional delivery				

Do learners have some background knowledge or skills that are needed to start the e-learning course?

❑   Yes

❑   No

❑   Not applicable

❑   Other (specify)

Do learners have previous experience with e-learning?

❑   Yes

❑   No

❑   Not applicable

❑   Other (specify)

What kinds of expectations do learners have concerning instructional delivery?

Goal Analysis

Does the course survey students before instruction begins in order to identify what they expect to learn or gain from the course?

❑   Yes

❑   No

❑   Not applicable
❑   Other (specify)

Are any of the following important aspects of an instructional goal considered in establishing each goal in the course?
❑   The learners
❑   The learning context
❑   The tools and technologies available to learners
❑   Other (specify)

How relevant is the instructional goal of the course to the learners?
❑   Highly relevant
❑   Moderately relevant
❑   Not relevant at all
❑   Not applicable
❑   Other (specify)

Are the course goal(s) approved by appropriate officials within the institution?
❑   Yes
❑   No
❑   Not applicable
❑   Other (specify)

Are there adequate resources (e.g., personnel, time, etc.) to develop e-learning lessons for the proposed course goal(s)?
❑   Yes
❑   No
❑   Not applicable
❑   Other (specify)

Check the appropriate course structure format for e-learning?
❑   Course → Unit → Lesson
❑   Other
❑   Not applicable

Outline the course structure below:

Course Name	Unit Name	Lesson Name
(For example) 101. Instructional design (ID)	(For example) 101.1 Introduction to ID	(For example) 101.1.1 Components of Systems approach Model
101.	101.1	101.1.1
		101.1.2
	101.2	101.2.1
		101.2.2
	101.3	101.3.1
		101.3.2
	101.4	101.4.1
		101.4.2

Length of learning units (check appropriate option):

❑ Less than 10 minutes

❑ 10 – 20 minutes

❑ 21 – 30 minutes

❑ 31 – 40 minutes

❑ 41 – 50 minutes

❑ Self-paced (depends on individual's progress)

❑ Instructor led

❑ Other

❑ NA

Does the course provide the following? (check all that apply):

Goals and Objectives	Yes	No	NA	Other
Course Goals				
Course Objectives				
Unit/Chapter Goals				
Unit/Chapter Objectives				
Lesson Goals				
Lesson Objectives				
Other (specify)				

Are clear learning outcomes specified in the course?
- ❑   Yes
- ❑   No
- ❑   Not applicable

Does the course provide clear descriptions of what capabilities learners will possess, what they should know or be able to do after completing the course?
- ❑   Yes
- ❑   No
- ❑   Not applicable

Are all required lesson objectives identified (that must be learned to achieve the lesson goal)?
- ❑   Yes
- ❑   No
- ❑   Not applicable

If appropriate, check all that apply for lesson objectives. Are they
- ❑   Measurable
- ❑   Achievable
- ❑   Not applicable
- ❑   Other (specify)

Does the course inform learners what they must do to achieve the objectives?
- ❑   Yes
- ❑   No
- ❑   Not applicable

Are course assignments, reports and discussions flexible enough to accommodate students' own learning goals?
- ❑   Yes
- ❑   No
- ❑   Not applicable

Do the course goals and objectives include the skills covered in similar courses taught in other institutions?

❑   Yes

❑   No

❑   Not applicable

Does the course review the prerequisite skills necessary for learning the skills of the course?

❑   Yes

❑   No

❑   None needed

❑   Not applicable

Media Analysis

Check if the course can use a variety of delivery media for its various lessons/units. Check all that apply:

Lesson	Delivery Media					
	Internet	CD-ROM	DVD	Print-Based Materials	Face-to-Face Class	Other
1.						
2.						
3.						
4.						
5.			/			
6.						

Which presentation modes does the course use? (check all that apply):

❑   Text

❑   Graphics

❑   Audio

❑   Video

❑   Animation

❑   Not applicable

❑   Other (specify below)

◀ How effective was the mixture of multimedia attributes in creating a rich environment for active learning?

❑   Very effective
❑   Moderately effective
❑   Not effective
❑   Not applicable
❑   Other

Does the course exploit the flexibility of the hypertext/hypermedia (e.g., hyperlinks) environment of the Web?

❑   Yes
❑   No
❑   Not applicable

Is the course content appropriately matched to the method of delivery?

❑   Yes
❑   No
❑   Not applicable

If *no*, describe below:

Design Approach

What type of content does the course deal with?

❑   Well structured
❑   Ill structured
❑   Combination of both (check the appropriate option from below)
    ❑   About equal percent of well structured or ill structured contents
    ❑   More well-structured than ill-structured contents
    ❑   More ill-structured than well-structured contents
    ❑   Other (specify)

Check the appropriate pedagogical philosophy for domain types: (check all apply)

Domain Type	Pedagogical Philosophy			
	Instructivist	Constructivist	Eclectic or Combination	Other (specify)
Well-structured				
Ill-structured				

Check the instructor's role (check all that apply):

- ❑ Domain expert
- ❑ Facilitator
- ❑ Coach
- ❑ Mentor
- ❑ Eclectic
- ❑ Not sure
- ❑ Other (specify)

Does the course allow for the instructor to serve as facilitator?

- ❑ Yes
- ❑ No
- ❑ Not applicable

If *yes*, how/where does facilitation occur? Can it occur in environments using any or all of the following Internet tools? (check all that apply):

- ❑ E mail
- ❑ Mailing list
- ❑ Online discussion forum
- ❑ Chat
- ❑ Audio conference
- ❑ Video conference
- ❑ Virtual classroom
- ❑ Other (specify)

What is the learner's role?

❑   Passive: A recipient of information

❑   Active: Active participant in creating knowledge from within

❑   Combination

❑   Not Applicable

❑   Not sure

❑   Other (specify below)

Does the course provide metacognition support by including annotations on online documents or resources?

❑   Yes

❑   No

❑   Not applicable

If *yes*, please describe how problems are presented and solved:

Please check the relevant control of learning activities used in the course.

❑   Student-centered (students control their own learning activities)

❑   Program-centered (students follows a structured environment)

❑   Combination of both

❑   Not sure

Is the course designed to support students to become independent distance learners?

❑   Yes

❑   No

❑   Not applicable

If *yes*, please specify how:

If the course allows students to have some control over the material to be learned, check all that apply:

❑   Students choose topics for course projects

❑   Students write up discussion questions

❑   Students select the path to navigate through instructions

❑   Students negotiate learning goals

❑   Select working group

❑   Students negotiate evaluation criteria

❑   Students have flexible due dates

❑   Other (specify)

If the course material is to some degree controlled by the program, check all that apply:

❑   Program determines the course's topics

❑   Program imposes the course's structure

❑   Program dictates the path through the instruction

❑   Program prescribes the learning goals

❑   Instructor forms teams for group projects

❑   Instructor leads discussions

❑   Instructor schedules exam dates

❑   Other (specify)

Organization

Does the course provide clear directions of what learners should do at every stage of the course?

❑   Yes

❑   No

❑   Not applicable

Does the course provide a sense of continuity for the learners (i.e., each unit of the lesson builds on the previous unit where appropriate)?

❑   Yes

❑   No

❑   Not applicable

Are the course materials organized in a manner appropriate to the apparent philosophical approach (instructivist, constructivist, or combination of both, etc.)?

❑   Yes, course material are organized following an instructivist instructional design approach

❑   Yes, course material are organized following a constructivist instructional design approach

❑   Yes, course material are organized following a combination of both instructivist and constructivist instructional design approach

❑   No

❑   Not applicable

❑   Other (specify)

Is there a clear and apparent sequence or structure to the information?

❑   Yes

❑   No

❑   Not applicable

Whenever appropriate, is the material grouped ("chunked") effectively?

❑   Yes

❑   No

❑   Not applicable

Does the course provide summaries of key points of the instruction?

❑   Yes

❑   No

❑   Not applicable

Learning Strategies[3]

Does the course have *online presentation(s)*?

❑   Yes

❑   No

❑   Not applicable

If *yes*, how effective were *online presentations* (check all that apply)?

Role of Individual	Performance Level				
	Excellent	Good	Fair	Poor	NA
Instructor					
Guest Speaker					
Students					
Other (specify)					

Are any of the following multimedia components, Internet tools, and supplementary materials used in *presentations*? (check all that apply):

I. Multimedia components
☐   Text
☐   Graphics
☐   Audio
☐   Animation
☐   Video
☐   Other (specify)

II. Internet tools
☐   E-mail
☐   Mailing lists
☐   Newsgroups
☐   Bulletin boards
☐   Chat
☐   Messaging
☐   Multi-user dialogues
☐   Computer conferencing
☐   Links to outside Websites
☐   Other (specify)

III. Supplementary materials
☐   CD-ROM
☐   DVD

❑   Videotape
❑   eBook
❑   Print (books/articles)
❑   Other (specify)

IV. Other (specify below)

Evaluate the *instructional* (e.g., learning related) and *technical* (e.g., bandwidth, file size, production quality, connectivity, etc.) effectiveness of the multimedia components, Internet tools, and supplementary materials in *presentations*.(check all that apply):

	Instructional Effectiveness					Technical Effectiveness				
	Excellent	*Good*	*Fair*	*Poor*	*NA*	*Excellent*	*Good*	*Fair*	*Poor*	*NA*
Multimedia										
Text										
Graphics										
Photographs										
Audio										
Narration										
Animation										
Video										
Other (specify)										
Internet tools										
E-mail										
Mailing lists										
Newsgroups										
Bulletin boards										
Chat										
Messaging										
Multi-user dialogues										
Computer conferencing										
Outside Website links										
Other (specify)										
Supplementary										
CD-ROM										
DVD										
Videotape										
eBook										
Print (books/articles)										
Other (specify)										

Does the course have virtual *exhibits*?

❑   Yes

❑   No

❑   Not applicable

     If *yes*, how effective were virtual *exhibits* used in the course?

     ❑   Very effective

     ❑   Moderately effective

     ❑   Not effective

     ❑   Other (specify)

Are all visuals and objects used in the digital *exhibits* organized with a clear description?

❑   Yes

❑   No

❑   Not applicable

❑   Other (specify)

Are any of the following multimedia components, Internet tools, and supplementary materials used in *exhibits*? (check all that apply):

I. Multimedia components

❑   Text

❑   Graphics

❑   Audio

❑   Animation

❑   Video

❑   Other (specify)

II. Internet tools

❑   E-mail

❑   Mailing lists

❑   Newsgroups

❑   Bulletin boards

❑   Chat
❑   Messaging
❑   Multi-User dialogues
❑   Computer conferencing
❑   Links to outside Websites
❑   Other (specify)

III. Supplementary materials

❑   CD-ROM
❑   DVD
❑   Videotape
❑   eBook
❑   Print (books, articles, etc.)
❑   Other (specify)

IV. Other (specify below)

Evaluate the *instructional* (e.g., learning related) and *technical* (e.g., bandwidth, file size, production quality, connectivity, etc.) effectiveness of the multimedia components, Internet tools, and supplementary materials in the instructional *exhibits*. (check all that apply):

	Instructional Effectiveness					Technical Effectiveness				
	Excellent	Good	Fair	Poor	NA	Excellent	Good	Fair	Poor	NA
Multimedia										
Text										
Graphics										
Photographs										
Audio										
Narration										
Animation										
Video										
Other (specify)										
Internet tools										
E-mail										
Mailing lists										
Newsgroups										
Bulletin boards										
Chat										
Messaging										
Multi-user dialogues										
Computer conferencing										

(continued from previous page)

Outside Website links												
Other (specify)												
Supplementary												
CD-ROM												
DVD												
Videotape												
eBook												
Print (books/articles)												
Other (specify)												

Does the course provide online *demonstration* sessions?
- ❑ Yes
- ❑ No
- ❑ Not applicable

    If *yes*, how effective were the *demonstration* sessions?
- ❑ Very effective
- ❑ Moderately effective
- ❑ Not effective
- ❑ Other (specify)

Are any of the following multimedia components, Internet tools, and supplementary materials used in the instructional *demonstrations*? (check all that apply):

I. Multimedia components
- ❑ Text
- ❑ Graphics
- ❑ Audio
- ❑ Animation
- ❑ Video
- ❑ Other (specify)

II. Internet tools
- ❑ E-mail
- ❑ Mailing lists
- ❑ Newsgroups

- ❑ Bulletin boards
- ❑ Chat
- ❑ Messaging
- ❑ Multi-user dialogues (MUDs)
- ❑ Computer conferencing
- ❑ Links to outside Websites
- ❑ Other (specify)

III. Supplementary materials

- ❑ CD-ROM
- ❑ DVD
- ❑ Videotape
- ❑ eBook
- ❑ Print (books, articles, etc.)
- ❑ Other (specify)

IV. Other (specify below)

Evaluate the *instructional* (e.g., learning related) and *technical* (e.g., band-width, file size, production quality, connectivity, etc.) effectiveness of the multimedia components, Internet tools, and supplementary materials in the *demonstration* sessions. (check all that apply):

	Instructional Effectiveness					Technical Effectiveness				
	Excellent	Good	Fair	Poor	NA	Excellent	Good	Fair	Poor	NA
Multimedia										
Text										
Graphics										
Photographs										
Audio										
Narration										
Animation										
Video										
Other (specify)										
Internet tools										
E-mail										
Mailing lists										
Newsgroups										
Bulletin boards										
Chat										
Messaging										
Multi-user dialogues										

(continued from previous page)

Computer Conferencing										
Outside Website links										
Other (specify)										
Supplementary										
CD-ROM										
DVD										
Videotape										
eBook										
Print (books/articles)										
Other (specify)										

Does the course provide online *drill and practice* sessions?

❑ Yes

❑ No

❑ Not applicable

If *yes*, how effective were the *drill and practice* sessions?

❑ Very effective

❑ Moderately effective

❑ Not effective

❑ Other (specify)

Are any of the following Multimedia components, Internet tools, supplementary materials used in *drill and practice*? (check all that apply):

I. Multimedia components

❑ Text

❑ Graphics

❑ Audio

❑ Animation

❑ Video

❑ Other (specify)

II. Internet tools

❑ E-mail

❑ Mailing lists

- ❑ Newsgroups
- ❑ Bulletin boards
- ❑ Chat
- ❑ Messaging
- ❑ Multi-user dialogues (MUDs)
- ❑ Computer conferencing
- ❑ Links to outside Websites
- ❑ Other (specify)

III. Supplementary materials
- ❑ CD-ROM
- ❑ DVD
- ❑ Videotape
- ❑ eBook
- ❑ Print (books, articles, etc.)
- ❑ Other (specify)

IV. Other (specify below)

Evaluate the *instructional* (e.g., learning related) and *technical* (e.g., bandwidth, file size, production quality, connectivity, etc.) effectiveness of the multimedia components, Internet tools, and supplementary materials in the *drill and practice* sessions. (check all that apply):

	Instructional Effectiveness					Technical Effectiveness				
	Excellent	Good	Fair	Poor	NA	Excellent	Good	Fair	Poor	NA
Multimedia										
Text										
Graphics										
Photographs										
Audio										
Narration										
Animation										
Video										
Other (specify)										
Internet tools										
E-mail										
Mailing lists										
Newsgroups										
Bulletin boards										

Continued from previous page

Chat											
Messaging											
Multi-user dialogues											
Computer conferencing											
Outside Website links											
Other (specify)											
Supplementary											
CD-ROM											
DVD											
Videotape											
eBook											
Print (books/articles)											
Other (specify)											

Does the course provide online *tutorial* sessions?

❑   Yes

❑   No

❑   Not applicable

If *yes*, how effective were the *tutorial* sessions?

❑   Very effective

❑   Moderately effective

❑   Not effective

❑   Other (specify)

Are any of the following multimedia components, Internet tools, and supplementary materials used in the *tutorials*? (check all that apply):

I. Multimedia components

❑   Text

❑   Graphics

❑   Audio

❑   Animation

❑   Video

❑   Other (specify)

II. Internet tools

- ❑ E-mail
- ❑ Mailing lists
- ❑ Newsgroups
- ❑ Bulletin boards
- ❑ Chat
- ❑ Messaging
- ❑ Multi-user dialogues (MUDs)
- ❑ Computer conferencing
- ❑ Links to outside Websites
- ❑ Other (specify)

III. Supplementary materials

- ❑ CD-ROM
- ❑ DVD
- ❑ Videotape
- ❑ eBook
- ❑ Print (books/articles)
- ❑ Other (specify)

IV. Other (specify below)

Evaluate the *instructional* (e.g., learning related) and *technical* (e.g., bandwidth, file size, production quality, connectivity, etc.) effectiveness of the multimedia components, Internet tools, and supplementary materials in the *tutorials*. (check all that apply):

	Instructional Effectiveness					Technical Effectiveness				
	Excellent	*Good*	*Fair*	*Poor*	*NA*	*Excellent*	*Good*	*Fair*	*Poor*	*NA*
Multimedia										
Text										
Graphics										
Photographs										
Audio										
Narration										
Animation										
Video										
Other (specify)										
Internet tools										
E-mail										
Mailing lists										
Newsgroups										
Bulletin boards										
Chat										
Messaging										
Multi-user dialogues										
Computer conferencing										
Outside Website links										
Other (specify)										
Supplementary										
CD-ROM										
DVD										
Videotape										
eBook										
Print (books/articles)										
Other (specify)										

Does the course use any *story-telling* technique?
- ❏   Yes
- ❏   No
- ❏   Not applicable

If *yes*, how effective were the *storytelling* techniques?
- ❏   Very effective
- ❏   Moderately effective
- ❏   Not effective
- ❏   Other (specify)

Are any of the following multimedia components, Internet tools, and supplementary materials used in the *storytelling*? (check all that apply):

I. Multimedia components
- ☐   Text
- ☐   Graphics
- ☐   Audio
- ☐   Animation
- ☐   Video
- ☐   Other (specify)

II. Internet tools
- ☐   E-mail
- ☐   Mailing lists
- ☐   Newsgroups
- ☐   Bulletin boards
- ☐   Chat
- ☐   Messaging
- ☐   Multi-user dialogues (MUDs)
- ☐   Computer conferencing
- ☐   Links to outside Websites
- ☐   Other (specify)

III. Supplementary materials
- ☐   CD-ROM
- ☐   DVD
- ☐   Videotape
- ☐   eBook
- ☐   Print (books/articles)
- ☐   Other (specify)

IV. Other (specify below)

Evaluate the *instructional* (e.g., learning related) and *technical* (e.g., bandwidth, file size, production quality, connectivity, etc.) effectiveness of the multimedia components, Internet tools, and supplementary materials in the instructional *storytelling*. (check all that apply):

	Instructional Effectiveness					Technical Effectiveness				
	Excellent	*Good*	*Fair*	*Poor*	*NA*	*Excellent*	*Good*	*Fair*	*Poor*	*NA*
Multimedia										
Text										
Graphics										
Photographs										
Audio										
Narration										
Animation										
Video										
Other (specify)										
Internet tools										
E-mail										
Mailing lists										
Newsgroups										
Bulletin boards										
Chat										
Messaging										
Multi-user dialogues										
Computer conferencing										
Outside Website links										
Other (specify)										
Supplementary										
CD-ROM										
DVD										
Videotape										
eBook										
Print (books/articles)										
Other (specify)										

Does the course use online *games*?

❑   Yes

❑   No

❑   Not applicable

If *yes*, how effective were the *games* sessions?

   ❑   Very effective

   ❑   Moderately effective

   ❑   Not effective

   ❑   Other (specify)

Are any of the following multimedia components, Internet tools, and supplementary materials used in the *games*? (check all that apply):

I. Multimedia components
- ❑ Text
- ❑ Graphics
- ❑ Audio
- ❑ Animation
- ❑ Video
- ❑ Other (specify)

II. Internet tools
- ❑ E-mail
- ❑ Mailing lists
- ❑ Newsgroups
- ❑ Bulletin boards
- ❑ Chat
- ❑ Messaging
- ❑ Multi-user dialogues (MUDs)
- ❑ Computer conferencing
- ❑ Links to outside Websites
- ❑ Other (specify)

III. Supplementary materials
- ❑ CD-ROM
- ❑ DVD
- ❑ Videotape
- ❑ eBook
- ❑ Print (books/articles)
- ❑ Other (specify)

IV. Other (specify below)

Evaluate the *instructional* (e.g., learning related) and *technical* (e.g., bandwidth, file size, production quality, connectivity, etc.) effectiveness of the multimedia components, Internet tools, and supplementary materials in the *game* sessions. (check all that apply):

	Instructional Effectiveness					Technical Effectiveness				
	Excellent	Good	Fair	Poor	NA	Excellent	Good	Fair	Poor	NA
Multimedia										
Text										
Graphics										
Photographs										
Audio										
Narration										
Animation										
Video										
Other (specify)										
Internet tools										
E-mail										
Mailing lists										
Newsgroups										
Bulletin boards										
Chat										
Messaging										
Multi-user dialogues										
Computer conferencing										
Outside Website links										
Other (specify)										
Supplementary										
CD-ROM										
DVD										
Videotape										
eBook										
Print (books/articles)										
Other (specify)										

Does the course use online *simulation*?

❑   Yes

❑   No

❑   Not applicable

> If *yes*, how effective were the *simulation* sessions?
>
> ❑   Very effective
>
> ❑   Moderately effective
>
> ❑   Not effective
>
> ❑   Other (specify)

Are any of the following multimedia components, Internet tools, and supplementary materials used in the *simulations*? (check all that apply):

I. Multimedia components
- ☐   Text
- ☐   Graphics
- ☐   Audio
- ☐   Animation
- ☐   Video
- ☐   Other (specify)

II. Internet tools
- ☐   E-mail
- ☐   Mailing lists
- ☐   Newsgroups
- ☐   Bulletin boards
- ☐   Chat
- ☐   Messaging
- ☐   Multi-user dialogues (MUDs)
- ☐   Computer conferencing
- ☐   Links to outside Websites
- ☐   Other (specify)

III. Supplementary materials
- ☐   CD-ROM
- ☐   DVD
- ☐   Videotape
- ☐   eBook
- ☐   Print (books/articles)
- ☐   Other (specify)

IV. Other (specify below)

Evaluate the *instructional* (e.g., learning related) and *technical* (e.g., bandwidth, file size, production quality, connectivity, etc.) effectiveness of the multimedia components, Internet tools, and supplementary materials in *simulations*. (check all that apply):

	Instructional Effectiveness					Technical Effectiveness				
	Excellent	*Good*	*Fair*	*Poor*	*NA*	*Excellent*	*Good*	*Fair*	*Poor*	*NA*
Multimedia										
Text										
Graphics										
Photographs										
Audio										
Narration										
Animation										
Video										
Other (specify)										
Internet tools										
E-mail										
Mailing lists										
Newsgroups										
Bulletin boards										
Chat										
Messaging										
Multi-user dialogues										
Computer conferencing										
Outside Website links										
Other (specify)										
Supplementary										
CD-ROM										
DVD										
Videotape										
eBook										
Print (books/articles)										
Other (specify)										

Does the course provide *role-playing* sessions?

❑   Yes

❑   No

❑   Not applicable

      If yes, were simulated *role portrayals* facilitated through:

❑   Multi-User Dialogue (MUD) environments, in which instructors create virtual space with a central theme, characters, and artifacts.

❑   Problem-based case studies
❑   Other (specify below)

How effective were the *role playing* sessions?
❑   Very effective
❑   Moderately effective
❑   Not effective
❑   Not applicable
❑   Other (specify below)

Are any of the following multimedia components, Internet tools, and supplementary materials used in the *role playing*? (check all that apply):

I. Multimedia components
❑   Text
❑   Graphics
❑   Audio
❑   Animation
❑   Video
❑   Other (specify)

II. Internet tools
❑   E-mail
❑   Mailing lists
❑   Newsgroups
❑   Bulletin boards
❑   Chat
❑   Messaging
❑   Multi-user dialogues (MUDs)
❑   Computer conferencing
❑   Links to outside Websites
❑   Other (specify)

III. Supplementary materials

- ❑ CD-ROM
- ❑ DVD
- ❑ Videotape
- ❑ eBook
- ❑ Print (books/articles)
- ❑ Other (specify)

IV. Other (specify below)

Evaluate the *instructional* (e.g., learning related) and *technical* (e.g., bandwidth, file size, production quality, connectivity, etc.) effectiveness of the multimedia components, Internet tools, and supplementary materials in the *role playing*. (check all that apply):

	Instructional Effectiveness					Technical Effectiveness				
	Excellent	Good	Fair	Poor	NA	Excellent	Good	Fair	Poor	NA
Multimedia										
Text										
Graphics										
Photographs										
Audio										
Narration										
Animation										
Video										
Other (specify)										
Internet tools										
E-mail										
Mailing lists										
Newsgroups										
Bulletin boards										
Chat										
Messaging										
Multi-user dialogues										
Computer conferencing										
Outside Website links										
Other (specify)										
Supplementary										
CD-ROM										
DVD										
Videotape										
eBook										
Print (books/articles)										
Other (specify)										

Does the course provide online *asynchronous discussion* sessions?
- ❑   Yes
- ❑   No
- ❑   Not applicable

If *yes*, how effective were the online *asynchronous discussion* sessions?
- ❑   Very effective
- ❑   Moderately effective
- ❑   Not effective
- ❑   Other (specify)

Does the course provide *online synchronous discussion* sessions?
- ❑   Yes
- ❑   No
- ❑   Not applicable

If *yes*, how effective were the *online synchronous discussion* sessions?
- ❑   Very effective
- ❑   Moderately effective
- ❑   Not effective
- ❑   Other (specify)

Are any of the following multimedia components, Internet tools, and supplementary materials used in the *discussions*? (check all that apply):

I. Multimedia components
- ❑   Text
- ❑   Graphics
- ❑   Audio
- ❑   Animation
- ❑   Video
- ❑   Other (specify)

II. Internet tools

- ❑ E-mail
- ❑ Mailing lists
- ❑ Newsgroups
- ❑ Bulletin boards
- ❑ Chat
- ❑ Messaging
- ❑ Multi-user dialogues (MUDs)
- ❑ Computer conferencing
- ❑ Links to outside Websites
- ❑ Other (specify)

III. Supplementary materials

- ❑ CD-ROM
- ❑ DVD
- ❑ Videotape
- ❑ eBook
- ❑ Print (books/articles)
- ❑ Other (specify)

IV. Other (specify below)

Evaluate the *instructional* (e.g., learning related) and *technical* (e.g., bandwidth, file size, production quality, connectivity, etc.) effectiveness of the multimedia components, Internet tools, and supplementary materials in the *discussion* sessions. (check all that apply):

	Instructional Effectiveness					Technical Effectiveness				
	Excellent	Good	Fair	Poor	NA	Excellent	Good	Fair	Poor	NA
Multimedia										
Text										
Graphics										
Photographs										
Audio										
Narration										
Animation										
Video										
Other (specify)										
Internet tools										
E-mail										
Mailing lists										
Newsgroups										
Bulletin boards										
Chat										
Messaging										
Multi-user dialogues										
Computer conferencing										
Outside Website links										
Other (specify)										
Supplementary										
CD-ROM										
DVD										
Videotape										
eBook										
Print (books/articles)										
Other (specify)										

Does the course instructor/facilitator post ground rules for the discussion forum?

❑   Yes

❑   No

❑   Not applicable

Does the course instructor/facilitator intervene when conflicts get personal in the discussion forum?

❑   Yes

❑   No

❑   Not applicable

Does the instructor or facilitator start the synchronous discussion session on time? (Note: if the facilitator is late, the learners may log off. In face-to-face classes learners may wait few a minutes or look for the instructor around the building, but online that may not happen. It should be noted that synchronous sessions sometimes may not start on time due to technical difficulties.)

❑   Yes
❑   No
❑   Not applicable

Are asynchronous discussion topics used in the course relevant to the goals and objectives of the course?

❑   Yes
❑   No
❑   Not applicable

Are synchronous discussion topics used in the course relevant to the goals and objectives of the course?

❑   Yes
❑   No
❑   Not applicable

Does the course require students to participate in scheduled online discussions?

❑   Yes
❑   No
❑   Not applicable

Does the course give students an opportunity to serve as online discussion leaders?

❑   Yes
❑   No
❑   Not applicable
❑   Other (specify)

Does the instructor/facilitator send private e-mails to those who are not participating in ongoing discussions?

❑   Yes

❑   No

❑   Not applicable

Does the instructor/facilitator send private e-mails to those whose messages appear to flame others on the class list?

❑   Yes

❑   No

❑   Not applicable

How does the instructor/facilitator communicate with individuals whose messages appear to flame others in the class list? (check all that apply):

❑   Private e-mail

❑   Telephone

❑   Online chat

❑   Online discussion

❑   Letter

❑   Other (specify)

Does the instructor/facilitator send private e-mails to those whose writings may be improved?

❑   Yes

❑   No

❑   Not applicable

Does the instructor/facilitator post encouraging messages on the list for students whose posts were thoughtful and relevant to the topic?

❑   Yes

❑   No

❑   Not applicable

Are learners advised to use a word processor in preparing their postings for discussion forums? (Note: I encourage my students to use the word processor for preparing their discussion form responses and save them on their hard drives. This way they can check spelling errors and grammar before posting it on the discussion forum. In the case of server failures, they can always retrieve their postings from their hard drives. However, some might argue that worrying about errors and typos can greatly inhibit students and waste their time.)

❑   Yes
❑   No
❑   Not applicable
❑   Other (specify)

Do students receive guidance on writing and online behavior on discussion forums?

❑   Yes
❑   No
❑   Not applicable

   If yes, check all that apply:
   ❑   How to write effective postings on discussion forums
   ❑   How to compose a response
   ❑   How to behave (netiquette) on a discussion forum
   ❑   Other

Are students encouraged to read and comment on each others' postings on online discussions?

❑   Yes
❑   No
❑   Not applicable

Does the instructor respond to students' postings on the discussion forum?

❑   Yes
❑   No
❑   Not applicable

If *yes*, check all that apply:

- ❑ Instructor responds to each student's posting.
- ❑ Instructor only responds to those postings where students ask for the instructor's attention.
- ❑ Instructor only responds to those postings to which a response seems appropriate, in the instructor's judgment.
- ❑ Instructor does not respond to students' postings on the discussion forums.

Does the instructor post online discussion topics on set dates (or at a scheduled time)?

- ❑ Yes
- ❑ No
- ❑ Not applicable

Are students required to submit discussion topics for class discussion?

- ❑ Yes
- ❑ No
- ❑ Recommended but not required
- ❑ Not applicable
- ❑ Other

Are students expected to assume a leadership role in moderating specific discussion topics at some time during the course?

- ❑ Yes
- ❑ No
- ❑ Not applicable

Does the instructor summarize and analyze the discussion at the end of each discussion topic?

- ❑ Yes
- ❑ No
- ❑ Not applicable

Does the instructor intervene appropriately when online discussions go in the wrong direction?

❑   Yes

❑   No

❑   Not applicable

❑   Other (specify)

Does the instructor/moderator encourage students to keep their posts brief and relevant to the discussion topic?

❑   Yes

❑   No

❑   Not applicable

Is the course discussion forum easy to use?

❑   Yes

❑   No

❑   Not applicable

Do students receive training in the use of the discussion forum?

❑   Yes

❑   No

❑   Not applicable

Does the course require or recommend that students subscribe to course relevant discussion forums?

❑   Yes

❑   No

❑   Not applicable

If *yes*, check all that apply:

Subscription	Required	Recommended
Class listserv		
Professional organizations' discussion lists		
Other (specify)		

Does the course instructor (or facilitator) signal the end of the on-going discussion by summarizing the discussion?

❑   Yes

❑   No

❑   Not applicable

Is the instructor (or facilitator) sensitive about potential information overload from the large flow of text generated from a discussion forum?

❑   Yes

❑   No

❑   Not applicable

   If *yes*, any preventive measures considered (please specify)

Does the course have a system of archiving synchronous discussions? (Note: This type of archive will be useful for students who cannot participate in live chats or who missed the live online discussion sessions. There is software that allows both voice and chat to be archived: http://www.horizonlive.com)

❑   Yes

❑   No

❑   Not applicable

Do the synchronous online discussion sessions provide for breaks (e.g., lunch breaks and periodical breaks)?

❑   Yes

❑   No

❑   Not applicable

   If *yes*, are they time zone sensitive?

How many participants are allowed to chat at the same time in synchronous environments? (It can be difficult to create effective live discussion sessions with too many learners actively participating.)

❑   Less than 10
❑   10 – 20
❑   21 – 30
❑   31 – 40
❑   41 – 50
❑   Not applicable
❑   Other

Are learners expected to do any asynchronous homework assignments before participating in a synchronous online discussion session?

❑   Yes
❑   No
❑   Not applicable

Are learners expected to have any specific materials in front of them during synchronous online discussion sessions?

❑   Yes
❑   No
❑   Not applicable

    If *yes*, check all that apply:
    ❑   Reading materials
    ❑   Calculator
    ❑   PowerPoint slides
    ❑   Notebook
    ❑   Not applicable
    ❑   Other (specify)

Does the course incorporate *interaction* as an instructional method?

❑   Yes
❑   No
❑   Not applicable

If *yes*, how effective were the *interactive* sessions?

❑   Very effective

❑   Moderately effective

❑   Not effective

❑   Other (specify)

Are any of the following multimedia components, Internet tools, and supplementary materials used in the *interactive sessions*? (check all that apply):

I. Multimedia components

❑   Text

❑   Graphics

❑   Audio

❑   Animation

❑   Video

❑   Other (specify)

II. Internet tools

❑   E-mail

❑   Mailing lists

❑   Newsgroups

❑   Bulletin boards

❑   Chat

❑   Messaging

❑   Multi-user dialogues (MUDs)

❑   Computer conferencing

❑   Links to outside Websites

❑   Other (specify)

III. Supplementary materials

❑   CD-ROM

❑   DVD

❑   Videotape

- ❑   eBook
- ❑   Print (books/articles)
- ❑   Other (specify)

IV. Other (specify below)

Evaluate the *instructional* (e.g., learning related) and *technical* (e.g., bandwidth, file size, production quality, connectivity, etc.) effectiveness of the multimedia components, Internet tools, and supplementary materials in the *discussion* sessions. (check all that apply):

	Instructional Effectiveness					Technical Effectiveness				
	Excellent	*Good*	*Fair*	*Poor*	*NA*	*Excellent*	*Good*	*Fair*	*Poor*	*NA*
Multimedia										
Text										
Graphics										
Photographs										
Audio										
Narration										
Animation										
Video										
Other (specify)										
Internet tools										
E-mail										
Mailing lists										
Newsgroups										
Bulletin boards										
Chat										
Messaging										
Multi-user dialogues										
Computer conferencing										
Outside Website links										
Other (specify)										
Supplementary										
CD-ROM										
DVD										
Videotape										
eBook										
Print (books/articles)										
Other (specify)										

Does the course encourage students to make comments about each other's assignments in the online discussion forum?

❑   Yes

❑   No

❑   Not applicable

Does the course encourage students to set up their own peer study groups?

❑   Yes

❑   No

❑   Not applicable

Is learner-learner interaction encouraged in the course?

❑   Yes

❑   No

❑   Not applicable

Does the course support interactions through the use of any of the following (check all that apply)?

❑   Peer evaluation

❑   Help sessions

❑   Collaborative projects

❑   Online study groups

❑   Not applicable

❑   Other (specify)

Is the course interactive?

❑   Yes

❑   No

❑   Not applicable

If *yes*, check all that apply:

❑   Among students?

❑   Between students and teacher(s)?

❑   With online resources?

How effective were the interactions?

☐   Very effective

☐   Moderately effective

☐   Not effective

☐   Not applicable

Does the course incorporate *modeling* as an instructional method?

☐   Yes

☐   No

☐   Not applicable

If *yes*, *modeling* is facilitated by:

☐   modeling behavior in electronic communication environments

☐   providing samples of relevant coursework

☐   providing guidance for interactions in simulated environments such as MUDs (Multi-User Dialogues)

☐   Other (specify below)

How effective were the *modeling* sessions?

☐   Very effective

☐   Moderately effective

☐   Not effective

☐   Not applicable

☐   Other (specify)

Are any of the following multimedia components, Internet tools, and supplementary materials used in the *modeling*? (check all that apply):

I. Multimedia components

☐   Text

☐   Graphics

☐   Audio

☐   Animation

❏   Video
❏   Other (specify)

II. Internet tools
❏   E-mail
❏   Mailing lists
❏   Newsgroups
❏   Bulletin boards
❏   Chat
❏   Messaging
❏   Multi-user dialogues (MUDs)
❏   Computer conferencing
❏   Links to outside Websites
❏   Other (specify)

III. Supplementary materials
❏   CD-ROM
❏   DVD
❏   eBook
❏   Print (books/articles)
❏   Other (specify)

IV. Other (specify below)

Evaluate the *instructional* (e.g., learning related) and *technical* (e.g., bandwidth, file size, production quality, connectivity, etc.) effectiveness of the multimedia components, Internet tools, and supplementary materials in the *modeling* sessions. (check all that apply):

	Instructional Effectiveness					Technical Effectiveness				
	Excellent	*Good*	*Fair*	*Poor*	*NA*	*Excellent*	*Good*	*Fair*	*Poor*	*NA*
Multimedia										
Text										
Graphics										
Photographs										
Audio										
Narration										
Animation										
Video										
Other (specify)										
Internet tools										
E-mail										
Mailing lists										
Newsgroups										
Bulletin boards										
Chat										
Messaging										
Multi-user dialogues										
Computer conferencing										
Outside Website links										
Other (specify)										
Supplementary										
CD-ROM										
DVD										
eBook										
Print (books/articles)										
Other (specify)										

Does the course use the instructional method of *facilitation* by providing guidance to students, directing discussion, suggesting possible resources, fielding questions, etc?

❑   Yes

❑   No

❑   Not applicable

If *yes*, please check all that apply:

Through asynchronous communication tools such as:

❑   E-mail

- ❏ Discussion forums
- ❏ Newsgroups
- ❏ Bulletin boards
- ❏ Web-based threaded discussions
- ❏ Not applicable
- ❏ Other (specify below)

Through synchronous communication tools such as:

- ❏ Chat
- ❏ Multi-user dialogues (MUDs)
- ❏ Audio conferencing
- ❏ Video conferencing
- ❏ Not applicable
- ❏ Other (specify below)

How effective were the *facilitation* sessions?

- ❏ Very effective
- ❏ Moderately effective
- ❏ Not effective
- ❏ Not applicable
- ❏ Other (specify)

Are any of the following multimedia components, Internet tools, and supplementary materials used in the *facilitation*? (check all that apply):

I. Multimedia components

- ❏ Text
- ❏ Graphics
- ❏ Audio
- ❏ Animation
- ❏ Video
- ❏ Other (specify)

II. Internet tools

- ❑ E-mail
- ❑ Mailing lists
- ❑ Newsgroups
- ❑ Bulletin boards
- ❑ Chat
- ❑ Messaging
- ❑ Multi-user dialogues (MUDs)
- ❑ Computer conferencing
- ❑ Links to outside Websites
- ❑ Other (specify)

III. Supplementary materials

- ❑ CD-ROM
- ❑ DVD
- ❑ Videotape
- ❑ E-book
- ❑ Print (books/articles)
- ❑ Other (specify)

IV. Other (specify below)

Evaluate the *instructional* (e.g., learning related) and *technical* (e.g., bandwidth, file size, production quality, connectivity, etc.) effectiveness of the multimedia components, Internet tools, and supplementary materials used in the service of course *facilitation*. (check all that apply):

	Instructional Effectiveness					Technical Effectiveness				
	Excellent	*Good*	*Fair*	*Poor*	*NA*	*Excellent*	*Good*	*Fair*	*Poor*	*NA*
Multimedia										
Text										
Graphics										
Photographs										
Audio										
Narration										
Animation										
Video										
Other (specify)										

Continued from previous page

Internet tools								
E-mail								
Mailing lists								
Newsgroups								
Bulletin boards								
Chat								
Messaging								
Multi-user dialogues								
Computer conferencing								
Outside Website links								
Other (specify)								
Supplementary								
CD-ROM								
DVD								
Videotape								
eBook								
Print (books/articles)								
Other (specify)								

Does the facilitator help learners focus on relevant issues in the discussion forum?

❑   Yes

❑   No

❑   Not applicable

Does the facilitator encourage learners to ask questions?

❑   Yes

❑   No

❑   Not applicable

Does the facilitator arouse interest and curiosity among learners?

❑   Yes

❑   No

❑   Not applicable

Does the facilitator encourage learners to elaborate their responses on issues discussed in the discussion forum?

❑   Yes

❑   No

❑   Not applicable

Does the facilitator encourage learners to reflect and self-evaluate?

❑   Yes

❑   No

❑   Not applicable

Does the facilitator provide a list of experts with whom learners can communicate via e mail to solicit expert opinions on issues related to their course projects?

❑   Yes

❑   No

❑   Not applicable

❑   Other (specify)

Does the facilitator provide customized responses for individual inquiries?

❑   Yes

❑   No

❑   Not applicable

Does the course provide a list of Frequently Asked Questions (FAQs) to handle questions that are asked over and over again?

❑   Yes

❑   No

❑   Not applicable

Does the course direct learners to explore external sites where they can analyze and compare materials? (Note: Such exploratory activities allow learners to make the materials relevant to their own needs and increase their motivation level.)

❑   Yes

❑   No

❑   Not applicable

Indicate the facilitator's level of involvement in facilitating online learning activities throughout the course?

❑   High-level involvement

❑   Mid-level involvement

❑   Low-level involvement
❑   Not applicable

Does the course promote *inside collaboration* by providing a supportive environment for asking questions, clarifying directions, suggesting or contributing resources, and class members working on joint projects?

❑   Yes
❑   No
❑   Not applicable

If *yes*, please check all that apply:

Through asynchronous communication tools such as:
❑   E-mail
❑   Discussion forums
❑   Newsgroups
❑   Bulletin boards
❑   Web-based threaded discussions
❑   Collaborative work tools that allow for shared screens
❑   Not applicable
❑   Other (specify below)

Through synchronous communication tools such as:
❑   Chat room
❑   Multi-user dialogues (MUDs)
❑   Computer conferencing
❑   Other (specify below)

How effective were the *inside collaboration* techniques?

❑   Very effective
❑   Moderately effective
❑   Not effective
❑   Not applicable
❑   Other (specify)

Does the course promote *outside collaboration* by involving external personnel and resources (speakers, guest lecturers, web sites, etc.) to participate in course activities?

❑   Yes

❑   No

❑   Not applicable

If *yes*, please check all that apply:

Through asynchronous communication tools such as:

    ❑   E-mail

    ❑   Discussion forum

    ❑   Newsgroups

    ❑   Bulletin boards

    ❑   Other (specify below)

Through synchronous communication tools such as:

❑   Chat

❑   Multi-user dialogues (MUDs)

❑   Computer conferencing

❑   Other (specify below)

Does the course have guest speakers? (Note: The course instructor or facilitator should ask learners to prepare their questions in advance and limit the number of questions so that the guest is not overwhelmed with questions.)

❑   Yes

❑   No

❑   Not applicable

If *yes*, please check all that apply:

    ❑   Information is provided for the number of times the guest speaker(s) will be available for synchronous discussion

    ❑   Information is provided for the period that the guest speaker(s) will be available for synchronous discussion

❏   Information about the guest speakers' contribution is clearly
    indicated

❏   Other (specify)

How effective were the outside collaboration techniques?

❏   Very effective

❏   Moderately effective

❏   Not effective

❏   Not applicable

Are any of the following multimedia components, Internet tools, and supplementary materials used in the *collaborative* sessions? (check all that apply):

I. Multimedia components

❏   Text

❏   Graphics

❏   Audio

❏   Animation

❏   Video

❏   Other (specify)

II. Internet tools

❏   E-mail

❏   Mailing lists

❏   Newsgroups

❏   Bulletin boards

❏   Chat

❏   Messaging

❏   Multi-user dialogues (MUDs)

❏   Computer conferencing

❏   Links to outside Websites

❏   Other (specify)

III. Supplementary materials

❑   CD-ROM

❑   DVD

❑   Videotape

❑   eBook

❑   Print (books/articles)

❑   Other (specify)

IV. Other (specify below)

Evaluate the *instructional* (e.g., learning related) and *technical* (e.g., bandwidth, file size, production quality, connectivity, etc.) effectiveness of the multimedia components, Internet tools, and supplementary materials in *collaborative* sessions. (check all that apply):

	Instructional Effectiveness					Technical Effectiveness				
	Excellent	*Good*	*Fair*	*Poor*	*NA*	*Excellent*	*Good*	*Fair*	*Poor*	*NA*
Multimedia										
Text										
Graphics										
Photographs										
Audio										
Narration										
Animation										
Video										
Other (specify)										
Internet tools										
E-mail										
Mailing lists										
Newsgroups										
Bulletin boards										
Chat										
Messaging										
Multi-user dialogues										
Computer conferencing										
Outside Website links										
Other (specify)										
Supplementary										
CD-ROM										
DVD										
Videotape										
eBook										
Print (books/articles)										
Other (specify)										

Does the course use *debates* as instructional activities?

☐ Yes

☐ No

☐ Not applicable

If *yes*, how effective were the online *debate* sessions?

☐ Very effective

☐ Moderately effective

☐ Not effective

☐ Other (specify)

Are any of the following multimedia components, Internet tools, and supplementary materials used in the *debates*? (check all that apply):

I. Multimedia components

☐ Text

☐ Graphics

☐ Audio

☐ Animation

☐ Video

☐ Other (specify)

II. Internet tools

☐ E-mail

☐ Mailing lists

☐ Newsgroups

☐ Bulletin boards

☐ Chat

☐ Messaging

☐ Multi-user dialogues (MUDs)

☐ Computer conferencing

☐ Links to outside Websites

☐ Other (specify)

III. Supplementary materials

- ❑ CD-ROM
- ❑ DVD
- ❑ Videotape
- ❑ eBook
- ❑ Print (books/articles)
- ❑ Other (specify)

IV. Other (specify below)

Evaluate the *instructional* (e.g., learning related) and *technical* (e.g., bandwidth, file size, production quality, connectivity, etc.) effectiveness of the multimedia components, Internet tools, and supplementary materials in the *debate* sessions. (check all that apply):

	Instructional Effectiveness					Technical Effectiveness				
	Excellent	*Good*	*Fair*	*Poor*	*NA*	*Excellent*	*Good*	*Fair*	*Poor*	*NA*
Multimedia										
Text										
Graphics										
Photographs										
Audio										
Narration										
Animation										
Video										
Other (specify)										
Internet tools										
E-mail										
Mailing lists										
Newsgroups										
Bulletin boards										
Chat										
Messaging										
Multi-user dialogues										
Computer conferencing										
Outside Website links										
Other (specify)										
Supplementary										
CD-ROM										
DVD										
Videotape										
eBook										
Print (books/articles)										
Other (specify)										

Do learners receive any guidelines in any of the following critical elements of debates? Check all that apply?

- ❑ How to engage in an open, honest exchange of ideas
- ❑ How to engage in group interaction
- ❑ How to think critically
- ❑ How to express personal views effectively
- ❑ How to be tolerant
- ❑ How to resolve conflicts among debate participants
- ❑ Other (specify)

Does the course use virtual *field trips* as an instructional method?

- ❑ Yes
- ❑ No
- ❑ Not applicable

    If *yes*, how effective were the online *field trips*?

- ❑ Very effective
- ❑ Moderately effective
- ❑ Not effective
- ❑ Other (specify)

Does the course provide students with a travel agenda and timetable for their online field trip?

- ❑ Yes
- ❑ No
- ❑ Not applicable

Are any of the following multimedia components, Internet tools, and supplementary materials used in the *field trips*? (check all that apply):

I. Multimedia components

- ❑ Text
- ❑ Graphics
- ❑ Audio

- ❑ Animation
- ❑ Video
- ❑ Other (specify)

II. Internet tools

- ❑ E-mail
- ❑ Mailing lists
- ❑ Newsgroups
- ❑ Bulletin boards
- ❑ Chat
- ❑ Messaging
- ❑ Multi-user dialogues (MUDs)
- ❑ Computer conferencing
- ❑ Links to outside Websites
- ❑ Other (specify)

III. Supplementary materials

- ❑ CD-ROM
- ❑ DVD
- ❑ Videotape
- ❑ eBook
- ❑ Print (books/articles)
- ❑ Other (specify)

IV. Other (specify below)

Evaluate the *instructional* (e.g., learning related) and *technical* (e.g., bandwidth, file size, production quality, connectivity, etc.) effectiveness of the multimedia components, Internet tools, and supplementary materials in the *field trip* sessions. (check all that apply):

	Instructional Effectiveness					Technical Effectiveness				
	Excellent	Good	Fair	Poor	NA	Excellent	Good	Fair	Poor	NA
Multimedia										
Text										
Graphics										
Photographs										
Audio										
Narration										
Animation										
Video										
Other (specify)										
Internet tools										
E-mail										
Mailing lists										
Newsgroups										
Bulletin boards										
Chat										
Messaging										
Multi-user dialogues										
Computer conferencing										
Outside Website links										
Other (specify)										
Supplementary										
CD-ROM										
DVD										
Videotape										
eBook										
Print (books/articles)										
Other (specify)										

Does the course provide students with specific guidelines for what they should accomplish through their field trip experience?

❑   Yes

❑   No

❑   Not applicable

Does the course require students to submit reports about their field trip?

❑   Yes

❑   No

❑   Not applicable

Are students required to discuss their field trip experience on the discussion forum?

❑   Yes
❑   No
❑   Not applicable

Does the course use *apprenticeship* as an instructional method (i.e., guidance by an outside expert for a particular learning task)?

❑   Yes
❑   No
❑   Not applicable

If *yes*, please check all that apply:

Through asynchronous communication tools such as:

❑   E-mail
❑   Discussion forums
❑   Newsgroups
❑   Bulletin boards
❑   Web-based threaded discussions
❑   Not applicable
❑   Other (specify below)

Through synchronous communication tools such as:

❑   Chat
❑   Multi-user dialogues (MUDs)
❑   Computer conferencing
❑   Other (specify below)

How effective were the *apprenticeship* sessions?

❑   Very effective
❑   Moderately effective
❑   Not effective
❑   Not applicable
❑   Other (specify)

Are any of the following multimedia components, Internet tools, and supplementary materials used in the *apprenticeship* activities? (check all that apply):

I. Multimedia components
- ☐   Text
- ☐   Graphics
- ☐   Audio
- ☐   Animation
- ☐   Video
- ☐   Other (specify)

II. Internet tools
- ☐   E-mail
- ☐   Mailing lists
- ☐   Newsgroups
- ☐   Bulletin boards
- ☐   Chat
- ☐   Messaging
- ☐   Multi-user dialogues (MUDs)
- ☐   Computer conferencing
- ☐   Links to outside Websites
- ☐   Other (specify)

III. Supplementary materials
- ☐   CD-ROM
- ☐   DVD
- ☐   Videotape
- ☐   eBook
- ☐   Print (books/articles)
- ☐   Other (specify)

IV. Other (specify below)

Evaluate the *instructional* (e.g., learning related) and *technical* (e.g., bandwidth, file size, production quality, connectivity, etc.) effectiveness of the multimedia components, Internet tools, and supplementary materials used to create the *apprenticeship* sessions. (check all that apply):

	Instructional Effectiveness					Technical Effectiveness				
	Excellent	*Good*	*Fair*	*Poor*	*NA*	*Excellent*	*Good*	*Fair*	*Poor*	*NA*
Multimedia										
Text										
Graphics										
Photographs										
Audio										
Narration										
Animation										
Video										
Other (specify)										
Internet tools										
E-mail										
Mailing lists										
Newsgroups										
Bulletin boards										
Chat										
Messaging										
Multi-user dialogues										
Computer conferencing										
Outside Website links										
Other (specify)										
Supplementary										
CD-ROM										
DVD										
Videotape										
eBook										
Print (books/articles)										
Other (specify)										

Does the course use *case studies*?

❑   Yes

❑   No

❑   Not applicable

   If *yes*, how effective were the *case studies*?

   ❑   Very effective

   ❑   Moderately effective

   ❑   Not effective

   ❑   Other (specify)

Are any of the following multimedia components, Internet tools, and supplementary materials used in the *case studies*? (check all that apply):

I. Multimedia components
☐   Text
☐   Graphics
☐   Audio
☐   Animation
☐   Video
☐   Other (specify)

II. Internet tools
☐   E-mail
☐   Mailing lists
☐   Newsgroups
☐   Bulletin boards
☐   Chat
☐   Messaging
☐   Multi-user dialogues (MUDs)
☐   Computer conferencing
☐   Links to outside Websites
☐   Other (specify)

III. Supplementary materials
☐   CD-ROM
☐   DVD
☐   Videotape
☐   eBook
☐   Print (books/articles)
☐   Other (specify)

IV. Other (specify below)

Evaluate the *instructional* (e.g., learning related) and *technical* (e.g., bandwidth, file size, production quality, connectivity, etc.) effectiveness of the multimedia components, Internet tools, and supplementary materials in the *case study* sessions. (check all that apply):

	Instructional Effectiveness					Technical Effectiveness				
	Excellent	*Good*	*Fair*	*Poor*	*NA*	*Excellent*	*Good*	*Fair*	*Poor*	*NA*
Multimedia										
Text										
Graphics										
Photographs										
Audio										
Narration										
Animation										
Video										
Other (specify)										
Internet tools										
E-mail										
Mailing lists										
Newsgroups										
Bulletin boards										
Chat										
Messaging										
Multi-user dialogues										
Computer conferencing										
Outside Website links										
Other (specify)										
Supplementary										
CD-ROM										
DVD										
Videotape										
eBook										
Print (books/articles)										
Other (specify)										

Does the course provide activities through which learners can generate understandings of course content? (Note: *Generative learning* can be supported by many different learning strategies.)

❏   Yes

❏   No

❏   Not applicable

If *yes*, for a course or unit, check the *generative strategies* used (check all that apply):

❑   Demonstrate comprehension of the facts, concepts, etc.

❑   Make predictions

❑   Paraphrase

❑   Summarize

❑   Elaborate

❑   Make inferences

❑   Devise applications (uses)

❑   Create metaphors or analogies

❑   Think of examples

❑   Diagram or visualize the structure of the new content

❑   Other (specify)

Does the course present the learner with authentic problem-solving activities in which the learner must make decisions and experience consequences?

❑   Yes

❑   No

❑   Not applicable

If *yes*, please describe how problems are presented and solved:

How effective were the *generative learning* methods?

❑   Very effective

❑   Moderately effective

❑   Not effective

❑   Not applicable

❑   Other (specify)

Are any of the following multimedia components, Internet tools, and supplementary materials used in the *generative learning*? (check all that apply):

I. Multimedia components
- ❏   Text
- ❏   Graphics
- ❏   Audio
- ❏   Animation
- ❏   Video
- ❏   Other (specify)

II. Internet tools
- ❏   E-mail
- ❏   Mailing lists
- ❏   Newsgroups
- ❏   Bulletin boards
- ❏   Chat
- ❏   Messaging
- ❏   Multi-user dialogues (MUDs)
- ❏   Computer conferencing
- ❏   Links to outside Websites
- ❏   Other (specify)

III. Supplementary materials
- ❏   CD-ROM
- ❏   DVD
- ❏   Videotape
- ❏   eBook
- ❏   Print (books/articles)
- ❏   Other (specify)

IV. Other (specify below)

Evaluate the *instructional* (e.g., learning related) and *technical* (e.g., band-width, file size, production quality, connectivity, etc.) effectiveness of the multimedia components, Internet tools, and supplementary materials in any activities that involve *generative learning* sessions. (check all that apply):

	Instructional Effectiveness					Technical Effectiveness				
	Excellent	Good	Fair	Poor	NA	Excellent	Good	Fair	Poor	NA
Multimedia										
Text										
Graphics										
Photographs										
Audio										
Narration										
Animation										
Video										
Other (specify)										
Internet tools										
E-mail										
Mailing lists										
Newsgroups										
Bulletin boards										
Chat										
Messaging										
Multi-user dialogues										
Computer conferencing										
Outside Website links										
Other (specify)										
Supplementary										
CD-ROM										
DVD										
Videotape										
eBook										
Print (books/articles)										
Other (specify)										

Does the course stimulate recall of prior knowledge?
- ❑ Yes
- ❑ No
- ❑ Not applicable

Does the course incorporate *motivation* as an instructional method?
- ❑ Yes
- ❑ No
- ❑ Not applicable

If *yes*, how effective were the *motivation* sessions?
- ❑ Very effective
- ❑ Moderately effective
- ❑ Not effective
- ❑ Other (specify)

Are any of the following multimedia components, Internet tools, and supplementary materials used to *motivate* students? (check all that apply):

I. Multimedia components
- ❑ Text
- ❑ Graphics
- ❑ Audio
- ❑ Animation
- ❑ Video
- ❑ Other (specify)

II. Internet tools
- ❑ E-mail
- ❑ Mailing lists
- ❑ Newsgroups
- ❑ Bulletin boards
- ❑ Chat
- ❑ Messaging
- ❑ Multi-user dialogues (MUDs)
- ❑ Computer conferencing
- ❑ Links to outside Websites
- ❑ Other (specify)

III. Supplementary materials
- ❑ CD-ROM
- ❑ DVD
- ❑ Videotape

❑   eBook
❑   Print (books/articles)
❑   Other (specify)

IV. Other (specify below)

Evaluate the *instructional* (e.g., learning related) and *technical* (e.g., band-width, file size, production quality, connectivity, etc.) effectiveness of the multimedia components, Internet tools, and supplementary materials used in *motivating* students. (check all that apply):

	Instructional Effectiveness					Technical Effectiveness				
	Excellent	*Good*	*Fair*	*Poor*	*NA*	*Excellent*	*Good*	*Fair*	*Poor*	*NA*
Multimedia										
Text										
Graphics										
Photographs										
Audio										
Narration										
Animation										
Video										
Other (specify)										
Internet tools										
E-mail										
Mailing lists										
Newsgroups										
Bulletin boards										
Chat										
Messaging										
Multi-user dialogues										
Computer conferencing										
Outside Website links										
Other (specify)										
Supplementary										
CD-ROM										
DVD										
Videotape										
eBook										
Print (books/articles)										
Other (specify)										

Does the course address concern for learner dissonance or anxiety? (Note: Learners' anxiety can be caused by the conflict between their beginner role, their lack of experience with Internet learning technologies, and their view of traditional learning systems, as indicated by Aggarwal, 2000. It is always good to discuss learner dissonance issues during orientation or the introductory session of the course.)

❑   Yes

❑   No

❑   Not applicable

❑   Other (specify)

Does the course provide for motivational factors such as fantasy and challenge, where appropriate?

❑   Yes

❑   No

❑   Not applicable

Does the course consider the situational and topical interest factors of cognitive motivation?

❑   Yes

❑   No

❑   Not applicable

Does the course provide ways to help students who are unmotivated about e-learning?

❑   Yes

❑   No

❑   Not applicable

At the beginning, does the course set an appropriate tone/climate in order for students to feel comfortable in sharing their ideas and personal information?

❑   Yes

❑   No

❑   Not applicable

Do students receive ongoing feedback on their performance in the various learning activities?

❑   Yes

❑   No

❑   Not applicable

Does the course encourage students to actively participate and contribute in online learning activities?

❑   Yes

❑   No

❑   Not applicable

❑   Other (specify)

Does the course use real world examples for students to make connections between course material and their lives?

❑   Yes

❑   No

❑   Not applicable

Does the course provide students with choice (such as options or alternatives and a sense of control over the learning environment)?

❑   Yes

❑   No

❑   Not applicable

❑   Other (specify)

Does the course provide students with a variety of learning activities to keep them interested and attentive?

❑   Yes

❑   No

❑   Not applicable

❑   Other (specify)

Does the course use motivational factors such as surprise, novelty, and intrigue to keep students curious about online learning activities?

❑   Yes

❑   No

❑   Not applicable

❑   Other (specify)

Does the course encourage students to exchange ideas and provide feedback on each other's work?

❑   Yes

❑   No

❑   Not applicable

❑   Other (specify)

Does the course provide examples and non-examples of new concepts and principles for the learners to make comparisons?

❑   Yes

❑   No

❑   Not applicable

Identify appropriate methods for various lessons or units of the course. Check all that apply:

Strategy	Lesson Name	Content Description
Presentation		
Exhibits		
Demonstration		
Drill and Practice		
Tutorials		
Storytelling		
Games		
Simulations		
Role-playing		
Discussion		
Interaction		
Modeling		
Facilitation		
Collaboration		
Debate		
Field Trips		
Apprenticeship		
Case Studies		
Generative learning		
Motivation		
Other (specify)		

Endnote

[1]   Please note that the author maintains a resource Website entitled "E-Learning Methods and Strategies" at http://BooksToRead.com/elearning/strategies.htm which provides links to relevant Websites dealing with various methods and strategies included in this section.

Chapter 6

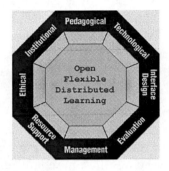

Ethical Issues

Ethical considerations of e-learning relate to social and political influence, cultural diversity, bias, geographical diversity, learner diversity, digital divide, etiquette, and the legal issues. The following is an outline of this chapter:

- Social and political influence
- Cultural diversity
- Bias
- Geographical diversity
- Learner diversity
- Digital divide
- Etiquette
- Legal issues

Social and Political Influence

Like any other innovative initiatives, e-learning projects can be subject to both social and political influence. "As we look at the distance learners, we must remember that these learners exist in a broad social context—a social context which can profoundly affect the success of the distance teaching-learning transaction" (Gibson, 1998, p. 113). E-learning projects funded by federal, state, or other sources may be subject to political influence by granting agencies. Moreover, power structure by various key players within the institution can influence the e-learning initiatives — they make or break a project! E-learning designers should pay attention to the cultural, social, and political influences that might affect their e-learning projects.

In some societies, political, and religious organizations may influence the decisions to include e-learning activities that they may feel inappropriate for their societies such as online learning collaborations or discussions with learners from societies differing in political ideologies and religious beliefs. In some countries, changes in government may influence the e-learning initiatives. In other societies, content or subject matter of e-learning courses/lessons can be influenced by political ideologies and religious beliefs. Also, the preferences for instructional methods and strategies can be influenced by political tradition. A memorandum on lifelong learning report from Norway's Ministry of Education released in July 2001 states that apprenticeship model is preferred by Norwegian political tradition:

In accordance with Norwegian political tradition, organizations representing employers and employees play an important part and an active role in the framing and implementation of education and training policies. This tradition of close co-operation occurs in both formal and informal ways. The social partners have traditionally had considerable influence on vocational training, especially in apprenticeship measures (http:// odin.dep.no/ufd/engelsk/publ/rapporter/014091-990010/index-hov004-b-n-a.html).

In summary, political, religious, and social traditions can impact e-learning. The restrictions may impinge on content, student activities, or the teaching strategies preferred.

Cultural Diversity

As a result of recent advances in distributed learning technologies, e-learning courses can be accessible to anyone in any part of the world. It is wonderful to be able to offer e-learning courses to learners around the globe with different social, cultural, economic, linguistic, and religious backgrounds. In designing e-learning environments, we should recognize the diversity of culture and learning styles in order to enhance learning for all (Sanchez & Gunawardena, 1998).

One of the difficult issues in e-learning is cross-cultural communication. Problems are encountered when at least one of the parties trying to exchange information is unaware of, or chooses to disregard, a significant difference in expectations concerning the relationships between communicators (Walls, 1993). Collis and Remmers (1997) remind us that we must be alert to the fact that there are substantial differences in interaction and communication beyond the actual words being said. In Bangladesh, the thumbs-up sign is used to challenge or disregard people, but to other cultures, that means you did well (Khan, 1999).

Designing e-learning for diverse learners is a challenging task. E-learning designers should try to be sensitive to cross-cultural communication issues. "In a global learning environment, designers, and developers must consider cultural sensitivities or risk loss of credibility" (Borman, 2001, p. 14). During the design process, they can ask individuals from various cultures to visit the course Web site and provide feedback. They can also post messages about cross-cultural issues on mailing lists whose members might be willing to share their experiences and point to appropriate resources.

Several authors provide guidance for cross-cultural issues related to e-learning (Collis & Remmers, 1997; Rice, Coleman, Shrader, Hall, Gibb & McBride, 2001). In some cultures, students use "sir" or "madam" as proper designation for the instructor. Instructors teaching students from around the globe should know that the use of proper designation for respected individuals is part of their culture. An online course offered globally should try to avoid using stories or examples that some of the learners may not be familiar with (e.g., using the US story of the Wizard of OZ may be foreign to many learners). Borman (2001) notes some examples of cultural nuances that course designers need to consider when building content:

- In Korea, funerals are typically bedecked in yellow. Yellow is the color of death and as such is a color to use with caution with Koreans.

- A barbecue sandwich and a beer might seem like an inviting lunch to you, but Indonesian and Malaysian people will find it offensive. Alcohol and pork are not served in those Muslim countries.

- Calling someone (even jokingly) a pig or dog in some Asian countries is inappropriate and dangerous.

Quigley (2002) notes an interesting cultural shock issue regarding the use of red color in China and in the USA:

Red is considered a lucky color in China. It's the traditional shade worn at weddings, the color of "hong baos", the money envelopes given on the Chinese New Year, and the predominant color of the national flag. How unlucky, then, that a US-based company planned to use red flags as warning signals in an e-learning course intended for Chinese students. Fortunately, someone noticed before it was too late.

Kathy Murrell of South Africa voiced her concern about the inappropriate use of sign or symbols in learning materials designed for South African audience by foreign designers. The following is an excerpt from her postings on the ITFORUM listserv:

In multiple choice questions the selected option was marked by a "tick" symbol (as apposed to a "cross"). I presume the American equivalent to be a check mark, but for most South Africans' this represents the mark next to a correct answer; it was most disconcerting to get an error or wrong message next to the mark and many students may not even bother to read the message as they would presume it to be the correct answer.

Well-recognized symbolic and iconic representations may not be understood by some ethnic groups within a country. Murrell's following ITFORUM listserv posting clearly informs us of the problems associated with symbols:

The internationally recognized symbol of an escape exit is taken by nonliterate Zulu speaking people to mean "do not go there your hands feet and head will be cut off", which makes sense when one looks closely at the image which has no neck, no wrist, and no ankle. This is exceptionally disturbing when one realises that these symbols were designed for foreign language speakers as well as people who are unable to read and write.

Bias

In e-learning, the content author's bias based on his or her position on issues should be carefully reviewed. An inclination toward a particular point of view can cause bias. However, sometimes it may be the case that an author is unaware of a bias. E-learning designers should work with content expert to eliminate any bias in the content. Learners should be informed about biases. For example, learners are encouraged to research topics which may be controversial. In their search for relevant articles on controversial issues, they may find Web sites that have been set up to express a particular viewpoint or ideology. Learners should be encouraged to find Web sites that present balanced viewpoint on controversial issues. For example, a history of Nepal might look very different if taught by a Pakistani, an Indian or faculty from China (Tom Abeles, DEOS listserve, Sub: Cross cultural issues Wed., 14 June 2000).

Geographical Diversity

Since e-learning can be offered to various geographical locations in the world, we should be very sensitive about participating students' locations, time zones, Internet accessibility, holidays, and so on.

The use of appropriate date and time conventions in e-learning provides orientation for a widely distributed group of students. I recommend the use of the full-text dating convention (e.g., March 1, 2003 instead of 01-03-2003) and UTC (Universal Coordinated Time) which is equivalent to GMT (Greenwich Mean Time) when arranging conference calls, online conferences, and other collaborative activities. The Web site http://www.collierad.com/coolebaytools/USTime.htm has a time conversion table for the regions of USA. The Web site http://www.worldtimeserver.com has an interactive time conversion option for countries around the world.

National and religious holidays may affect e-learning activities for geographically dispersed learners. Students may not feel comfortable submitting their assignments or participating in online activities during holidays. Instructors teaching such diverse populations should recognize holidays and travel time associated with those holidays and come up with a clear guidelines on such issues that are fair to all. For example, groups of students engaged in joint projects from different time zones may face difficulty in coordinating their schedules due to holidays and travel. Scheduled chat discussions may not work for learners coming from different time zones. In summary, all e-learning courses should be sensitive to holidays and diversity in geographical time zones.

Technical constraints may prevent some learners from doing certain e-learning activities. For example, electrical power outages, load shading, and circuit failure rates in different parts of the world may serve as barriers for learners' participation in scheduled synchronous learning activities and submit their assignments on time. E-learning designers should take these technical problems into account in scheduling synchronous learning activities and assignment due dates for geographically dispersed learners.

Economic constraints may also prevent learners in some regions of the world from doing certain e-learning activities completely online. For example, learners who pay "by the minute" Internet connection fees (or long-distance charge) will be affected by any online discussion activities requiring heavy connection hours. Alan Selig made the following comments about the connection fees issue on the eModerators discussion list on November 3, 2000:

I agree that composition time is a key value for anyone working in a "second" language. I do have a question regarding webbed discussion boards, however. Does the student have to remain connected while reading from a discussion board, or can he/she download the submissions and then log off the net? I've only used e-mail lists. I like them because a participant can read the submissions without having to remain connected to the Internet. If that is not a possibility for webbed discussion boards, then it would be an additional expense for anyone who is paying "by the minute" connection fees or long-distance charges. That's probably not a very big issue in the US, but I know it is in some other countries. A discussion group should probably know the locations of its members and whether a "read on the Web" format would be an additional financial burden.

Learner Diversity

An e-learning system should be designed to accommodate different learning styles and the needs of individuals with disabilities, including senior citizens whose physical faculties (e.g., hearing) are impaired; in the US alone, it has been estimated that there are more than 30 million people with disabilities — inborn, acquired, and temporary.

Every learner has his/her own style for meaningfully gathering and organizing information for learning. Illinois Online Network (ION) Web site at http://illinois.online.uillinois.edu/IONresources/instructionaldesign/learningstyles.html discuses some common learning styles for online learning: visual/verbal (prefers to read information), visual/nonverbal (uses graphics or diagrams to represent information), auditory/verbal (prefers to listen to information), and tactile/ kinesthetic (prefers physical hands-on experiences). Catherine Jester at Diablo Valley College developed an online quiz on learning style titled "A Learning Style Survey for College." The result of the quiz gives scores on the following categories: visual/nonverbal, visual/verbal, auditory and kinesthetic (http:// www.metamath.com//multiple/multiple_choice_questions.cgi). E-learning designers can use a similar quiz to get an overall picture of their target audience and their learning styles.

The diversity of student population in online courses poses a challenge to the instructor to develop a lecture (or lecturette) to meet the needs of a potentially wider range of students in background, experience, and expertise (Fullmer-Umari, 2000). When it comes to online synchronous and asynchronous discussions, a non-native speaker of the discussion forum may encounter difficulty in expressing clearly in his/her writings which may present him/her as different to other participants. The instructor or discussion moderator should be sensitive in situations like this. They should find innovative ways to involve diverse learners in e-learning activities. At the beginning of the course, I recommend the following activities for the instructor:

- The instructor should ask all students in the course to post their brief biographies, encourage them to read each other's posting, and continue dialoging with each other throughout the course duration.

- The instructor can send a welcome note to the class indicating that the class should be proud of having a wider range of students in background, experience, and expertise. Learners in the class, through mutual respect and patience, can create a great learning experience. The class should be patient with non-native speakers as they may take longer time to compile and post their responses in discussion forums. In addition, the instructor may consider writing a personal note to non-native speakers encouraging them to take more time (if they need) to compose their responses for forum questions.

In designing online courses, multiple instructional strategies and activities that cater to various learning styles should be used. Williams and Peters (1997) noted that e-learning offers a better environment to accommodate flexibility in learning styles. An e-learning course offered to learners around the globe should consider using examples familiar to its target audience. Please note that we may not be able to create e-learning environments that will fit every type of learning style, but an understanding of the learning styles can help us provide alternatives whenever possible.

Digital Divide

In an information society, information accessibility is a critical issue which must be discussed in terms of the gap between the digital "haves" and "have nots," a gap expressed in the term "digital divide". Digital divide refers to the gap between those who have access to the Internet and other information technology and those who do not.

There are many reasons why a digital divide exists. The barriers are evident when individuals (1) are unable to afford Internet access from home, (2) find e-learning content that is difficult to comprehend, (3) find e-learning content that is not culturally-relevant. It is important to note that people with disabilities often confront more barriers in e-learning than others.

The digital divide can also exist among different geographic locations. In some countries, some areas have broadband or high speed Internet services and some areas have only dial-up connection via telephone. It seems that having access to the Internet does not mean that there is no digital divide. There is definitely a big difference between accessing Internet via high speed (i.e., T1 line, DSL, Cable modem, etc.) connection versus low speed dial-up connections.

Different Internet access speeds obviously create a digital divide among high and low speed Internet users. Sometimes the problem of digital divide can be caused by where one lives. If high speed Internet services are not available in some locations, then their residents have to rely on low speed dial-up connections. This is an example of the digital divide issue caused by geographical locations. Maybe in the near future, this divide will not be an issue for anyone to discuss. I brought it up here to present a situation that digital divide can be caused by business decisions by Internet service providers.

Can we eliminate digital divide? Whether or not we can permanently eliminate digital divide, we can definitely minimize the gap. To minimize the gap, communities, and governments around the globe should develop policies which should be followed by providers of technology infrastructure, hardware and software, Internet services, and e-learning contents. These policies should focus on how to make equal access to e-learning resources and materials available to wider population.

The Cybrarian Project at the Department for Education and Skills (DfES) in the United Kingdom is dedicated to assisting in decreasing the digital divide by facilitating access to the Internet and to learning opportunities for those who currently do not, or cannot, use the Internet because of a lack of skills or confidence or because of physical or cognitive disabilities (http://www.dfes.gov.uk/cybrarianproject/index.cfm). In the United States, the Web-Based Education Commission made the recommendation led by Senator John Kerry and Congressman Johnny Isakson to make powerful new Internet resources, especially broadband access, widely and equitably available and affordable for all learners. Kerry and Isakson (2000) state: "We call on federal and state governments to make the extension of broadband access for all learners a central goal of telecommunications policy" (p. 129).

Digital divide has been widely discussed in many online forums and publications. Discussion of issues related to the digital divide can be found at DigitalDivideNetwork.org. In writing the editorial comments on a special issue of *Educational Technology Review,* French (2002) states:

First, providing equal access to educational opportunities is simply the right thing to do. Vast amounts of valuable educational material are being gleaned from the Internet every day by students and professors alike. Just as physical accessibility is routine across campuses, virtual accommodations are just as necessary across the Internet. From this ethical perspective most individuals agree that education should be made accessible to everyone.

What can e-learning designers do to minimize the gap? In designing e-learning activities, digital divide issues should be considered to include the learners who are affected by this division. Since the loading speed on the Internet may vary with users' Internet connection speeds, e-learning designers should use multimedia elements that are essential to content. E-learning designers need to respect differences in bandwidth. Individuals with slow and unreliable Internet connections have to wait longer time to download large files. Since images and videos without text alternatives are inaccessible to learners who are visually impaired for any reason, the use of alternate text for all non-text elements is essential in this regard. (See *Accessibility* section in Chapter 7 for more information about the interface design issues for people with disabilities.)

Etiquette

An e-learning environment should have the guidelines for netiquette (network or Internet etiquette) especially when students post messages on discussion forums, newsgroups, and interact with others in the course via e-mail or instant messaging. Etiquette provides rules for maintaining civility in interactions and covers issues associated with considerate behavior. The etiquette promotes mutually respectful behavior in an online learning community.

Shea (1994) in her book Netiquette, presented 10 core rules of netiquette which can be used in e-learning courses. These core rules are discussed with practical examples at http://www.albion.com/netiquette/corerules.html:

- Remember the human
- Adhere to the same standards of behavior online that you follow in real life
- Know where you are in cyberspace
- Respect other people's time and bandwidth
- Make yourself look good online
- Share expert knowledge
- Help keep flame wars under control
- Respect other people's privacy
- Do not abuse your power
- Be forgiving of other people's mistakes

Participants in both synchronous and asynchronous communications should not personally attack others. Personal attacks not only disturb the learning process, but also discourage interactions and collaborations among participants.

In the fall of 2003, a discussion forum was held among students from the University of Texas at Brownsville, Texas A&M University, Kansas State University, and the University of Washington. The purpose of the forum was to create a collaborative learning experience on issues related to distance learning by implementing online student-to-student distance dialogue (Murphy, Khan, Knupfer & Cifuentes, 1997). Before setting up the discussion forum, we instructors met and selected discussion topics suited for the participating students. One of the instructors moderated the forum. Students from all four campuses participated in the discussion forum. It had a good start. However, after one of the students flamed another student, an individual who was neither a student nor an instructor supported the student who flamed, thereby jeopardizing the goal of the collaborative discussions forum. As a result, students slowly disengaged themselves.

After reflecting on what the instructors could have done to prevent the problem, they realized that they should have been more aggressive in imposing rules for maintaining civility in the forum.

To prevent problems in online discussions, I would recommend that e-learning courses set up orientation sessions for netiquette. Discussions are the heart of any e-learning course. All participants in asynchronous and synchronous discussions should be knowledgeable about netiquette rules, and follow them appropriately. Therefore, we should try our best to encourage students to meaningfully participate in knowledge sharing collaborative learning environments. The Web site at http://www.albion.com/netiquette/netiquiz.html hosts a quiz where one can test his or her knowledge about netiquette.

The use of special language such as emoticons (e.g., :-V for "shout"), abbreviations (e.g., IMHO for "in my humble opinion") and technical terms (e.g., logon) during online communications is not uncommon. However, both instructor and learners should be knowledgeable about the meaning and appropriate usage of them. The Web site at http://210.210.18.114/EnlightenmentorAreas/it/EI/Netiquette.htm provides basic emoticons that are commonly used in chat and e-mail.

Legal Issues

Institutions should develop e-learning policies and guidelines for legal matters such as privacy, plagiarism, and copyright issues at the very beginning of their e-learning initiatives. On October 3, 2002, the United States Senate passed the Technology, Education and Copyright Harmonization Act, commonly known as the "TEACH Act," outlining the use of copyrighted materials in the virtual classroom. The TEACH Act provides some much needed clarification and expansion of privileges for distance learning. These policies and guidelines should be followed by all stakeholder groups, including instructors, learners, and e-learning administrative and support services staff.

East Carolina University (ECU) has developed policies and guidelines for the content and appearance of documents and other subject matter contained on all Web pages. ECU's Policy on the World Wide Web can be found at: http://www.ecu.edu/webdev/policy.html. North Carolina State University provides an online tutorial on copyright ownership, copyright use, and plagiarism. This tutorial is designed to help the university faculty, staff, and students with questions concerning works created during employment or enrollment at the University. The tutorial is available at: http://www.lib.ncsu.edu/scc/tutorial/

Privacy

In a typical e-learning course, there can be numerous text dialogs generated from mailing lists or computer conferencing exchanges. These exchanges may contain participants' personal views and biases which they may not want the outside world to know. Considering the openness of the Web, it is not difficult for search engines to find these exchanges. Palloff and Pratt (1999) suggest that participants must know that their communications are not secure and that they must use good judgment about what they share. The following concern is voiced by Shawn Foley who posted his comments on the subject of "Ethical review for computer-conferencing records" on the eModerators listserve on December 8, 2000:

Who owns the data? If I take an online course, I add to the computer conferencing record. If you remove all identifiable information the data that remains is still "my" data. Whether it can be linked to me or not, I have a right to refuse the use of that data, since without me it would not exist. I see that data as partially my property, and unless I sign an agreement turning that property over to specific parties for specific purposes, the data should not be used for research purposes.

Institution should have privacy policy. The institution should clearly indicate to the students whether or not it will share their personal information and text dialogs to others. Both instructor and students should never publish or forward private e-mail message without permission, and this policy should be put into writing in the appropriate form and place. At the beginning of the course, students should be informed about the openness of the Web and privacy guidelines. Even during the online discussions, the instructor, tutor, or facilitator should remind students about the privacy issues.

Plagiarism

Plagiarism in e-learning can happen when one steals another person's writing and presents it as one's own. It is like using someone else's materials without crediting the source. With a search engine or specialized directory, anyone can find reliable (and unreliable) content on any subject on the Internet. Plagiarism in e-learning is easier because one can copy and paste materials from Internet sites with click of a mouse. Like many colleges and universities throughout the United States and across the globe, Jones International University contracts the services of TurnItIn.com, an online resource for monitoring the originality of student work (http://jiu-web-a.jonesinternational.edu/eprise/main/JIU/studentcenter/JIU_turnitin.html?banner=student).

Students should be cautioned about presenting someone else's work as their own. For example, when a student copies from other students' assignments (either partial or full copying) is considered plagiarism. E-learning courses should provide clear information regarding institution's plagiarism policies. Athabasca University in Canada imposes serious penalties if a case of plagiarism is substantiated. Athabasca University's "Student Code of Conduct and Right to Appeal" policies can be found at: http://www.athabascau.ca/calendar/02/conduct12.html.

In the design of the course assignments, we should think about ways that will make assignments more interesting, challenging, and personal. These assignments should not motivate students to steal from others; rather they should encourage learners to back up their ideas with others' point of views. Lynch (2002) provides the following tips for deterring plagiarism in e-learning:

- Make sure students understand that when they copy something from the Web it is the same as copying something from a book. Believe it or not, many students do not know this.

- Have students sign a contract of understanding and agreement not to plagiarize. This may not stop them, but it will make them think twice.

- Strictly enforce the rules when you discover plagiarism (e.g., a failing grade on that paper, dismissal from the school, etc.).

- Include a module on plagiarism in your online student orientation course. Have students work through examples where they are required to identify if the example is plagiarized and, if so, how the reference should have been cited.

- Provide an online class environment that requires a great deal of interaction in both formal and informal discussions, writing papers, and working with peers.

- Include at least one telephone discussion or a time when the instructor phones the students for a "check-up chat". Ask how the student is feeling about the class, question the student about recent difficult concepts, and offer assistance for learning those concepts better.

- If you use a final exam or competency exam that must be proctored, then work to obtain pre-approved acceptable proctors and sites well in advance (e.g., local college faculty, teaching and learning center personnel, librarians, military personnel, church pastors, etc.).

- Focus on field-based application of concepts. For example, when teaching a business communications course online, require students to give a presentation in a business environment and to have it critiqued by a manager. The critique form is then forwarded to the instructor. You need not grade them on the critique scores themselves, as you cannot control grading curves from one manager to another. However, you may elect to grade students' reflection papers describing their experience in giving the presentation, what they learned from the critique, and how they will improve in the future.

E-learning courses can easily use all of the above-mentioned tips to deter plagiarism. However, one could ask the question on phoning a student who lives outside of the instructor's country, "Who will pay for the international telephone charges?" The simple answer would be: if the institution is serious about quality education and preventing plagiarism, it should consider allocating some funds for long distance phone charges.

Is there a way to avoid telephone charges in e-learning? One possibility would be to use Internet Phone (assuming both the student and instructor have audio capabilities on their computers). Lynch states: "I was using Internet Phone as early as 1997 over 28.8K lines and later 56K lines. It did not work well over 28.8, but at 56K it is pretty good" (Maggie McVay Lynch, personal communication, February 25, 2003). She recognizes that some countries pay high phone charges just to connect to the Internet, so it might still be prohibitive.

As more and more institutions offer online courses to geographically diverse learners, we will continually learn what works and what does not work. However, we already know what really works, as Lynch states, "My experience has been that if you follow the rules about learner-centered assignments and discussion board postings that are reflective of the learner's individual experience, you get a sense of the learner's writing style and it is easy to spot when that style changes drastically. That is really the best way to approach the problem (and the use of the software tools) (Maggie McVay Lynch, personal communication, February 25, 2003).

The Web site "Tempted by the Web: Teaching Students to Avoid Plagiarism and Detection Tools for Teachers" at http://web.pdx.edu/~mmlynch/plagiarism2.htm provides resources and information on plagiarism.

Copyright

Chapter 2 discusses intellectual property rights issues relate to ownership of learning materials developed by faculty members. Content authors, instructors, tutors, facilitators, guest speakers, and students should be knowledgeable about copyright issues pertaining to e-learning and obtain copyright permissions wherever appropriate (see Table 1). If an instructor decides to make a journal article available to students on the Web, he or she should get the permission from the journal. Therefore, to cite or use any electronic information on the Internet (e.g., postings from discussion forums, e-mail messages, etc.) for a project, it is important to contact the author of that particular electronic postings to confirm its authenticity and validity, and ask for permission to use his/her electronic postings (see Table 2).

An interesting legal interpretation of whether the use of educational materials on the Internet constitutes copyright infringement can be found at http://

Table 1. Permission for copyrighted graphic

Copyrighted Graphic	Sample E-mail Message to Use the Graphic
  I receive e-mail messages from individuals asking for permission to use the above copyrighted (© Badrul Khan) graphic which is available at the following site: BooksToRead.com/framework I usually I grant permission for the educational use of the graphic.	From: "Gail Brooks" <gbrooks@mwc.edu> Date: Tue, 25 Feb 2003 20:14:59 -0500 To: <khanb@BooksToRead.com> Subject: question Dear Dr. Khan, I am a PhD student at George Mason University. I'm currently taking a course on distance learning. One of our projects involves picking a distance learning site and evaluating it. We found your e-learning framework a wonderful tool for evaluation. We plan to use it in our presentation. We would also like to include your elearning framework graphic in our presentation. We wanted to get your permission to do so. This is a presentation for educational purposes only. We will, of course, site you as owner of the graphic and cite you as the source for the framework we plan to use. Your response is greatly appreciated. Best Regards, Gail Brooks

Table 2. Permission for citing discussion posting

From: "Badrul H Khan" <khanb@BooksToRead.com>
Date: Thu, 27 Feb 2003 18:20:10 -0500 (EST)
Subject: citing your discussion forum posting in my book
To: Tom Abeles <tabeles@tmn.com>

Hello Mr. Tom Abeles:

My name is Badrul Khan. I am currently authoring a book entitled "E-learning Strategies" (http://www.BooksToRead.com/elearning/). In the book, I would like to cite your following quote which you posted on the DEOS discussion forum (subject: cross cultural issues, June 14, 2000):

"a history of Nepal might look very different if taught by a Pakistani, an Indian or faculty from China"

Thanks for your PROMPT reply to CONFIRM it.

Best wishes,
Badrul H. Khan

www.ivanhoffman.com/onlinefair.html. The University of Texas System Crash Course in Copyright is a tutorial program which is available for faculty to use to learn copyright basics, especially in the distance learning context (http://www.utsystem.edu/ogc/intellectualproperty/cprtindx.htm).

Question to Consider

Can you think of any e-learning related ethical issues not covered in this chapter?

Activity

1. Using Internet search engines, locate at least one article that discusses any of the following ethical issues; and analyze its usefulness in e-learning programs.

    •   Digital divide

    •   Etiquette

    •   Legal issues

    •   Privacy

    •   Plagiarism

    •   Copyright

2. Locate an online program and review it from the perspectives of ethical considerations checklist items at the end of Chapter 6.

References

Borman, R. (2001). Training a global audience: Avoid making cross-cultural boo-boos. *e-learning, 2*(7), 13-15.

Collis, B. & Remmers, E. (1997). The WWW in education: Issues related to cross-cultural communication and interaction. In B.H. Khan (Ed.), *Web-based instruction,* (pp. 85-92). Englewood Cliffs, NJ: Educational Technology Publications.

French, D. (2002). Accessibility...an integral part of online learning. *Educational Technology Review*. Retrieved February 17, 2003, from *http://www.aace.org/pubs/etr/issue2/french-ed.cfm*

Fullmer-Umari, M. (2000). Getting Ready. In K.W. White & B.H. Weight (Eds.). *The Online Teaching Guide*. Needham Height, MA: Allyn & Bacon.

Gibson, C.C. (1998). Distance learner in context. In C. C. Gibson (Ed.), *Distance learners in higher education*. Madison, Wisconsin: Atwood Publishing.

Khan, B.H. (1999). Interviewed by Debra Donston for an article titled "From the trenches: Distributed learning is high priority," *PCWEEK, 16*(46), 134.

Lynch, M. (2002). *The online educator: A guide to creating the virtual classroom*. London & New York: Routledge.

Murphy, K., Khan, B.H., Knupfer, N. & Cifuentes L. (1997). Implementing online student-to-student distance dialogue: Adding depth to local course offerings. Paper presented at the *Annual Meeting of the Association for Educational Communications and Technology (AECT)*, Albuquerque, NM.

Palloff, R.M. & Pratt, K. (1999). *Building learning communities in cyberspace*. San Francisco, CA: Jossey-Bass Publications.

Quigley, A. (2002). Culture shock: Overseas e-learning markets require. *eLearn Magazine*. Retrieved January 24, 2003, from *http://www.elearnmag.org/subpage/sub_page.cfm?article_pk=4061& page_number_nb=1& title= FEATURE%20STORY*

Rice, J., Coleman, M.D., Shrader, V.E., Hall, J.P., Gibb, S.A. & McBride, R.H. (2000). Developing Web-based training for global corporate community. In B.H. Khan (Ed.), *Web-based training,* (pp. 191-202). Englewood Cliffs, NJ: Educational Technology Publications.

Sanchez, I. & Gunawardena, C.N. (1998). Understanding and supporting the culturally diverse distance learner. In C.C. Gibson (Ed.), *Distance learners in higher education*. Madison, WI: Atwood Publishing.

Walls, J. (1993). Global networking for local development: Task force and relationship focus in cross-cultural communication. In L. Harasim (Ed.), *Global networks: Computers and international communication*. Cambridge, MA: MIT Press.

Williams, V. & Peters, K. (1997). Faculty incentives for the preparation of Web-based Instruction. In B.H. Khan (Ed,), *Web-based instruction,* (pp. 107-110). Englewood Cliffs, NJ: Educational Technology Publications.

Ethical Checklist

Social and Political Influence

Does the institution have to get approval from any external entities (that can serve as political barriers) to implement its e-learning?

❑ Yes

❑ No

❑ Not applicable

If *yes*, please list the entities:

Does the course designer need internal approval from any authorities within the institution for certain e-learning content and activities?

❑ Yes

❑ No

❑ Not applicable

If *yes*, please list e-learning content types, e-learning activities and the approving authorities:

Is there a social/political preference for any particular instructional method? (Note: for example, the apprenticeship model is preferred by Norwegian political tradition.)

❑ Yes

❑ No

❑ Not applicable

If *yes*, please list the e-learning strategies most preferred:

Cultural Diversity

To improve cross-cultural verbal communication and avoid misunderstanding, does the course make an effort to reduce or avoid the use of jargon, idioms, humor, acronyms, and ambiguous words, terms and content? (Note: We should avoid using jokes or comments that can be misinterpreted and misunderstood by some.)

❑   Yes

❑   No

❑   Not applicable

> If *yes*, does the course have or link to resource site(s) where interpretations of cross cultural jargon and idioms are available?
>
> ❑   Yes
>
> ❑   No
>
> ❑   Not applicable

To improve visual communication, is the course sensitive to the use of navigational icons or images? (Note: For example, Reeves & Reeves in 1997 noted that a pointing hand icon to indicate direction would violate a cultural taboo in certain African cultures by representing a dismembered body part. Also, a pointing finger that indicates a hyperlink would be problematic too. A right arrow for the next page may mean previous page for Arabic and Hebrew language speakers as they read from left to right).

❑   Yes

❑   No

❑   Not applicable

Does the course use the full name for acronyms used in the body of the text?

❑   Yes

❑   No

❑   Not applicable

If *yes*, are the acronyms used for terms globally understood? (Note: Acronyms for terms such as identification numbers used in different parts of the world can be confusing to learners. Many countries of the world have identification or record keeping mechanisms for their citizens. For example, the United States government uses the acronym SSN for Social Security Number; the Canadian government uses the acronym SIN for Social Insurance Number, etc. Therefore, it will be problematic when a course offered by an US institution asks for a SSN number from   non-US students.

❑   Yes

❑   No

❑   Not applicable

Could a student find the course to be discriminatory? (Note: It is difficult to judge on what may offend one person, but not another.)

❑   Yes

❑   No

❑   Not applicable

If *yes*, please describe:

Is the course culturally sensitive?

❑   Yes

❑   No

❑   Not applicable

❑   Not sure

Is the course sensitive to learners who come from an oral culture?

❑   Yes

❑   No

❑   Not applicable

Does the course promote cross-cultural interaction among students and instructor(s)?

❑   Yes

❑   No

❑   Not applicable

Is the course offered in multilingual format? (Note: Text in buttons or icons is harder to change. Hornett in an article written in 2000 entitled "Culturally Competent" advised us not to include text in graphics for e-learning content with the potential for being translated into other languages.)

❑   Yes

❑   No

❑   Not applicable

If *yes*, indicate the names of languages:

How does the course address cultural diversity from a learning perspective? (check all that apply):

❑   Course is tailored to specific cultures

❑   Course is designed to be culturally neutral

❑   Not applicable

Does the course use any icons, images, graphics, etc. which may have offensive meanings for learners of various cultures?

❑   Yes

❑   No

❑   Not applicable

❑   Not sure

Does the course vary the representation of concepts to allow for a multicultural audience?

❑   Yes

❑   No

❑   Not applicable

Does the course use terms or words that may not be used by the worldwide audience? (Note: People use the term "sidewalk" in the US and "pavement/footpath" in the UK. When such a term is needed, we should include both forms for a diverse audience, such as "students should use the sidewalk [or pavement] rather than trample the grass." The Website http://www.eurotexte.fr/translation/tips_brit_vs_amer.shtml provides some of the differences between American and British English.)

❑   Yes

❑   No

❑   Not applicable

Does the course use signs or symbols that may not be used by a worldwide audience?

❑   Yes

❑   No

❑   Not applicable

Does the course use symbolic and iconic representations that are not always commonly understood within one country? (Note: In South Africa, Kathy Murrell noted on the ITFORM listserve that the internationally recognized symbol of an escape exit is taken by nonliterate Zulu speaking people to mean "don't go there your hands, feet, and head will be cut off" - which makes sense when one looks closely at the image which has no neck, no wrist and no ankle.)

❑   Yes

❑   No

❑   Not applicable

Does the course use culture-specific analogies, metaphors, or expressions?

(Note: For example, "Be sure to save your work frequently, remember *a stitch in time saves nine.*")

❑   Yes

❑   No

❑   Not applicable

Bias

Is the course sensitive to the biases of the authors of the content?

❑   Yes

❑   No

❑   Not applicable

Does the course present more than one viewpoint on controversial issues?

❑   Yes

❑   No

❑   Not applicable

Does the course designer try to eliminate any bias in the course content?

❑   Yes

❑   No

❑   Not applicable

Is the course content bias-free?

❑   Yes

❑   No

❑   Not applicable

Geographical Diversity

Is the course offered to geographically diverse populations?

❑   Yes

❑   No

❑   Not applicable

If *yes*, is the course sensitive about students from different time-zones (e.g. synchronous communications are scheduled at reasonable times for all time zones represented)?

❑   Yes

❑   No

❑   Not applicable

For assignment due dates, is the instructor sensitive to national and religious holidays observed by students (not observed by the instructor)?

❑   Yes

❑   No

❑   Not applicable

Is the instructor sensitive about scheduling synchronous learning activities (such as chat) during national and religious holidays observed by students (not observed by the instructor)?

❑   Yes

❑   No

❑   Not applicable

Check if any of following issues are considered in scheduling online-learning activities? (check all that apply):

❑   Electrical power outages, load shading and circuit failure in some parts of the world may affect learners participating in online synchronous activities

❑   Electrical power outages, load shading and circuit failure in some parts of the world may prevent learners from submitting assignments on time

❑   Not applicable

❑   Other

Are the Internet connection fees a deterrent to participation in certain online activities by learners? (Note: Learners who pay "by the minute" connection fees, or long-distance charges may be affected by online discussion activities requiring long connections times.)

❑   Yes

❑   No

❑   Not applicable

❑   Other

Learner Diversity

Does the institution conduct a survey to assess the learning style of the target population?

❑   Yes

❑   No

❑   Not applicable

❑   Other

Does the course provide flexibility to accommodate diverse learning styles?

❑   Yes

❑   No

❑   Not applicable

Is the course designed to have patience for learners who adapt to the distributed learning environment more slowly than others?

❑   Yes

❑   No

❑   Not applicable

Does the course allow students to remain anonymous during online discussions?

❑   Yes

❑   No

❑   Not applicable

Does the course foster mutual respect, tolerance, and trust? (Note: Such an environment depends on what the instructor and all learning support staff do during the course.)

❑   Yes

❑   No

❑   Not applicable

Does the course allow students to lurk during online *synchronous* discussions?
- ❏   Yes
- ❏   No
- ❏   Not applicable

Does the course allow students to lurk during online *asynchronous* discussions?
- ❏   Yes
- ❏   No
- ❏   Not applicable

Are participants required to assume roles or participate in scenarios that might be culturally and religiously offensive (Note: An e-learning course can ask learners to join in multi-person interactive worlds such as MOOs, MUDs and MUSHes to play and talk under a variety of personae. However, the topic and the roles that learners are asked to play may be culturally offensive. For instance, asking learners to play homosexual and bisexual characters in a MUD session may be ethically and religiously offensive to some learners.)
- ❏   Yes
- ❏   No
- ❏   Not applicable

Is the course designed to accommodate the needs of visually impaired learners?
- ❏   Yes
- ❏   No
- ❏   Not applicable

    If *yes*, check all that apply:
- ❏   Learners can use "text-to-reader" software to participate in the course
- ❏   Other (specify)

Does the course offer an audio version for visually impaired learners?
- ❏   Yes
- ❏   No
- ❏   Not applicable

Digital Divide

Is the digital divide issue considered in designing the e-learning content?

❑　Yes

❑　No

❑　Not applicable

 If *yes*, check all of the following measures that apply

 ❑ Only essential multimedia elements are used in the course to reduce bandwidth problem

 ❑ Multimedia elements (graphics, audio, video) are accompanied by text equivalents to be accessible by people with disabilities

 ❑ Other (specify)

Is the course sensitive to a diverse student population's accessibility to the Internet?

❑　Yes

❑　No

❑　Not applicable

Etiquette

Does the institution have etiquette guidelines?

❑　Yes

❑　No

❑　Not applicable

Does the course provide any guidance to learners on how to behave and post messages in online discussions so that their postings do not hurt others' feelings?

❑　Yes

❑　No

❑　Not applicable

If a student fails to follow the etiquette of the course more than one time, how does the instructor work with each student to promote compliance? (check all that apply):

❑   The student receives final notices with consequences
❑   The student is put on probation
❑   The student is penalize by lowering his/her grade or points
❑   The student is remove from the discussion forum
❑   Other
❑   Not applicable

Legal Issues

Does the course comply with the institution's policies and guidelines (if any) regarding all Web page development?

❑   Yes
❑   No
❑   Not applicable

Does the institution provide privacy policies and guidelines on online postings?

❑   Yes
❑   No
❑   Not applicable

> If *yes*, does the course comply with the institution's privacy policies and guidelines for online postings?
>
> ❑   Yes
> ❑   No
> ❑   Not applicable

Does the course provide ethics policies that outline rules, regulations, guidelines, and prohibitions?

❑   Yes
❑   No
❑   Not applicable

Does the course get previous students' permission to use their online discussions postings or any other data that belong to them?

❑   Yes

❑   No

❑   Not applicable

Does the institution store students' text dialogs generated from mailing lists or computer conferencing exchanges?

❑   Yes

❑   No

❑   Not applicable

If *yes*, does it release students' text dialogs to others? (check all that apply):

❑   Yes, it releases students' text dialogs to other with students' permission

❑   Yes, it releases students' text dialogs to other without students' permission

❑   No

❑   Not applicable

Does the course provide institutional policies and guidelines regarding fraudulent activities in course-related testing, assignments and projects?

❑   Yes

❑   No

❑   Not applicable

Does the course clearly inform students about the consequences (e.g., in terms of course grade and academic status) of any forms of plagiarism?

❑   Yes

❑   No

❑   Not applicable

If *yes*, check all that apply:
- ❑   Receive a failing grade in the course
- ❑   Receive a failing grade on that particular paper
- ❑   Dismissal from the institution
- ❑   Name shows up on the list of cheaters in the institution
- ❑   Institution shares the student's cheating record with other academic institutions
- ❑   Other

Does the course provide a mini lesson on plagiarism?
- ❑   Yes
- ❑   No
- ❑   Not applicable

If *yes*, check all that apply:
- ❑   Learners identify an example of plagiarism
- ❑   Learners are advised how to cite or give credit to the source
- ❑   Other

Does the course require students to sign an agreement on plagiarism?
- ❑   Yes
- ❑   No
- ❑   Not applicable

Does the institution have a legal office where faculty members can get answers to legal matters concerning online courses?
- ❑   Yes
- ❑   No
- ❑   Not applicable
- ❑   Other (specify)

Check if the institution provides training sessions with up-to-date information on copyright issues relevant to e-learning? (check all that apply):

Role of Individual	Training Session Format			
	Online	Face-to-Face	Other	Not Applicable
Learner				
Instructor (full-time)				
Instructor (part-time)				
Trainer				
Trainer Assistant				
Tutor				
Technical Support				
Help Desk				
Librarian				
Counselor				
Graduate Assistant				
Administrator				
Project Manager				
Instructional Designer				
Graphic Artist				
Programmer				
Multimedia Developers				
Other (specify)				

Does the course acquire permission to use copyrighted information and materials from appropriate copyright holders?

❑   Yes

❑   No

❑   Not applicable

Does the course get students' permission to post any of the following on the Web? (check all that apply):

Student Materials	Permission			
	Yes	*No*	*NA*	*Other*
Students' projects				
Students' Webfolios				
Students' photographs				
Students' email addresses				
Students' telephone numbers				
Students' mailing address				
Other				

Does the course provide information about institutional policies and guidelines about copyrights?

❑   Yes

❑   No

❑   Not applicable

❑   Other (specify)

Does the course provide appropriate information about copyright laws concerning learning activities on the Internet?

❑   Yes

❑   No

❑   Not applicable

Chapter 7

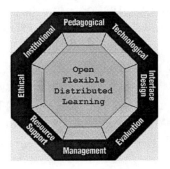

Interface
Design Issues

Interface design refers to the overall look and feel of an e-learning program (Hall, 1997). User interface design is the creation of a seamless integration of content and its organization, together with the navigational and interactive controls that learners use to work with the content (Jones & Farquhar, 1997). The design of an e-learning interface is critical because it determines how learners interact with the presented information (Brown, Milner & Ford, 2001). In an e-learning environment, all e-learning interfaces should be designed to accommodate the tasks of learners while they access information. Lohr (2000) states that learning interfaces are defined as those where communication cues take place between a learner and a learning system.

Lohr (1998) suggests three critical roles of the instructional interface: (1) to provide learner orientation to instructional content, (2) to provide navigational tools to access instructional content and instructional strategies, and (3) to provide feedback. Mayer's (1993) identification of selection, organization, and integration cognitive processes is a convenient way to emphasize the organization suggested by Lohr.

Selection is the process that takes place when the learner notices the important information and is able to isolate it from less important information. As seen in Table 1 (Lohr, in press), many of the orientation features of the interface address this function. Instructional interface designers should seek to help the learner notice the most salient information when first accessing instruction. For example, a clearly identified topic is an important piece of information that can set the stage for greater understanding.

Organization is the process where the learner is able to chunk or sequence information in a way that is meaningful. In interface design, organization elements include visual cues and features similar to a table of contents that help a learner mark where they are in the instruction. Navigation panes, menus, page or screen numbers, and links all become important organization features.

Table 1. Cognitive processes associated with anticipated learner questions

Cognitive processes involved	Anticipated learner questions (Lohr, 2000; Lohr, 2003)
Selection (noticing the critical information)	**Anticipating the following *orientation* questions:** • What is the topic of learning? • How do I begin learning? • What is the learning climate? • What is the breadth of this environment? • What, in general, is expected of me in this learning environment? • Do I feel comfortable welcome in this environment?
Organization (chunking or sequencing information)	**Anticipating the following *navigation* questions:** • What is the depth? • Where am I in this process? • Can I mark where I am? • How do I go back? • What do I do now? • What do I do next? • When am I finished? • How do I get out of this?
Integration (assimilating or accommodating information)	**Anticipating the following *instructional strategies access* questions:** • How do I interact with this instructional strategy? • Can I get more/less information? More or less examples? • Can I skip this information?
	Providing interactive feedback • Am I doing the right thing? • Am I right/wrong? • How did I end up here? • Can I undo what I just did? • Can I customize this?

Integration is the process where the learner is able to assimilate or accommodate new information into memory. Integration interface elements are those that facilitate access to and understanding of learning spaces. Buttons that link to extra help, pop-up windows that explain new terms, graphic organizers integrated into the interface, and space for practice and feedback help the learner integrate.

Table 1 shows a list of anticipated learner questions and how these questions relate to cognitive processes involved in accessing information.

This chapter is organized with the following outline to encompass various critical issues of instructional interfaces discussed above by Lohr:

- Page and site design
- Content design
- Navigation
- Accessibility
- Usability testing

Page and Site Design

Web page and site design in e-learning relates to the appearance and functionality of the screen. We have to make sure that all Web pages in a Web site are logically organized, easy to navigate, easily accessible, and usable by all users, including people with disabilities (for more information, see *Accessibility and Usability Testing* sections in this chapter).

Damarin (2000) states that whenever feasible, we should make e-learning materials available in multiple formats. This multiplicity gives students the fullest possible access to instructional materials. Users should be enabled to select the optimal interface structure, interacting with content by reading and typing, pointing and clicking, and using speech recognition and synthesis as desired. Damarin stresses that the Web documents should also be available for browsing, printing, and saving in various formats (PDF, HTML, word, text) whenever possible.

Content Design

The content has to do with a course's subject matter. Quality content is one of the most important determinants of Web usability (Nielsen, 2000). Nielsen provides guidelines for designing quality content. His guidelines encompass text density (i.e., keeping text short), copy editing (i.e., proofreading, spelling, and grammar checking), scannability (i.e., text layout that can be quickly understood), and so on. A good overview of content design by Nielsen can be found at http://www.useit.com/alertbox/9710a.html.

In the presentation of content, e-learning courses should always strive for clarity, style, and readability. Standard writing conventions such as grammar, capitalization, punctuation, usage, spelling, paragraphing, and so on should be used effectively for clarity, style, and readability. Content-relevant graphics (e.g., icons, buttons, pictures, images, etc.) and other multimedia components (e.g., audio, video, etc.) should be used appropriately to supplement textual contents. The writing style should be simple, clear, and direct, and must be appropriate for the reading level of target audience.

E-learning screen design should use good message design principles that focus on learner attention, perception, comprehension, retention, and retrieval (Morrison, Ross & O'Dell, 1995). Instructional illustrations Web site at http://dev.comet.ucar.edu/presentations/illustra/index.htm demonstrates how we can use pictures to help learners understand and remember information in ways that text and lecture often do not.

Navigation

Designing navigation in an e-learning should focus on enabling learners to move through the site with ease and reasonable speed. Clarity and consistent use of textual, graphic, and other organizational markers throughout the site can contribute to the ease of use and speed (Simich-Dudgeon, 1998). Due to lack of clarity and consistency, learners can become disoriented and, as a result, lose motivation.

Accessibility

Interface design in an open, flexible and distributed e-learning environment should focus on how to make it accessible to all potential students. If we are able to design a face to face classroom in a building to meet the needs of all students, including those with disabilities, then we should do the same thing for e-learning sites. However, designing e-learning for diverse learners in an open, flexible, and distributed learning environment poses new challenges.

In designing e-learning sites, we should be aware of various barriers that make e-learning materials inaccessible to some students. These barriers may be caused by technical and design issues.

Technical issues such as bandwidth issue may serve as accessibility barrier. For example, a large file in an e-learning course can take longer time to download or in some case it may be impossible to download. This is a bandwidth issue, which in turn serve as an accessibility barrier for users who do not have high speed Internet connection. Therefore, e-learning courses should be bandwidth efficient for all users.

Let us talk about design issues. To accommodate the needs of learners, especially people with disabilities, we have to identify situations in e-learning which may serve as barriers and then find ways to overcome these barriers. For example, images and videos without text alternatives are inaccessible to learners who are visually impaired for any reason. The use of alternate text for all non-text elements can be read aloud by software for synthesizing speech.

The World Wide Web Consortium (W3C) has developed a series of influential accessibility guidelines for Web design. W3C provides Web Content Accessibility Guidelines 1.0. These guidelines explain how to make Web content accessible to people with disabilities (http://www.w3.org/TR/1999/WAI-WEBCONTENT-19990505/).

Therefore, all e-learning materials should be designed based on Web accessibility guidelines. The Center for Applied Special Technology (http://bobby.cast.org) provides a Web-based tool called Bobby that analyzes Web pages for W3C compliance.

In designing e-learning materials, designers need to make sure that any multimedia elements are essential to content and accompanied by text equivalents.

Accessibility issues are taken seriously in many parts of the world. In some parts of the world accessibility can be a legal issue. For example, it is a law (Section 508) in the United States for all federal agencies to make their electronic and information technology accessible to people with disabilities. Therefore, all federally funded e-learning projects in the United States must be Section 508 compliant. It means that if an educational institution receives federal funds, then it must offer equitable access to technology for all students.

The following Web sites have information about how some countries are making guidelines for accessibility issues.

Regions	URL
USA	http://section508.gov/
Canada	http://www.cio-dpi.gc.ca/clf-upe/index_e.asp
European Union (EU)	http://europa.eu.int/information_society/topics/citizens/accessibility/index_en.htm
Australia	http://www.govonline.gov.au/projects/standards/accessibility.htm
General information	http://www.icdri.org

Also, textbooks and other learning materials should be available in structured electronic format such as HTML or XML to improve accessibility for people with disabilities. Mikhail Vaysbukh in an article titled "Leveling the textbook playing field for the print disabled," discusses the benefits of electronic form for people with print disabilities. "When structured data is read by a text-to-speech engine, for example, it can supply crucial information like what page or heading level you are on. It also helps generate a navigational structure, which enables the user to move from level to level and be able to recognize things like the start of the next exercise" (http://www.dclab.com/accessibility_whitepaper.asp).

When print materials such as textbooks are available in XML form, it helps e-learning programs to blend or integrate them effectively. An increasing number of states in the United States are requesting that textbooks and other learning materials are available in structured electronic format to improve accessibility for people with disabilities. Vaysbukh reports that "In early 2004, for example, Kentucky's state legislature will require all educational materials be in well-formed HTML or in XML files that meet accessibility requirements."

Usability Testing

Usability testing is a tool for improving interface design. Typically, users evaluate an e-learning program to make sure it is usable. Reeves and Carter (2001) categorize usability testing as follows: efficiency (i.e., cost and time saving), user satisfaction (i.e., ease of use, intuitiveness, visual appeal, etc.), and effectiveness (i.e., user retention over time). Guidelines for designing usable graphical user interfaces and Web pages can be found at: http://www.useit.com/. Nielsen (2000) recommends e-learning programs to perform international usability testing with users from a few countries in different parts of the world.

Question to Consider

Can you think of any e-learning related interface design issues not covered in this chapter?

Activity

1. Using Internet search engines, locate an article that covers any of the following interface design issues for online courses; and analyze the article from the perspectives of its usefulness in e-learning:

   • Navigation

   • Accessibility

   • Usability testing

2. Locate an online program and review its interface design aspects using the relevant interface design checklist items at the end of Chapter 7.

References

Brown, K.G., Milner, K.R. & Ford, J.K. (2000). Repurposing instructor-led training into Web-based training: A case study and lessons learned. In B.H. Khan (Ed.), *Web-based training,* (pp. 415-422). Englewood Cliffs, NJ: Educational Technology Publications.

Damarin, S.K. (2000). The 'digital divide' versus digital differences: Principles for equitable use of technology in education. *Educational Technology*, March-April, 17-22.

Hall, B. (1997). *Web-based training cookbook.* New York: Wiley.

Jones, M.G. & Farquhar, J.D. (1997). User interface design for Web-based instruction. In B.H. Khan (Ed.), *Web-based instruction,* (pp. 239-244). Englewood Cliffs, NJ: Educational Technology Publications.

Lohr, L. (in press). Instructional interface design for Web learning. In B.H. Khan (Ed.), *Flexible learning*. Englewood Cliffs, NJ: Educational Technology Publications.

Lohr, L. (2003). *Creating graphics for learning and performance: Lessons in visual literacy*. Columbus, Ohio: Prentice-Hall.

Lohr, L. (2000). Designing the instructional interface. *Computers in Human Behavior, 16*(2), 161-182.

Morrison, G., Ross, S. & O'Dell, J. (1995). Applications of research to the design of computer-based instruction. In G. Anglin (Ed.) *Instructional technology: Past, present, and future* (2nd ed.). Englewood, CO: Libraries Unlimited, Inc.

Nielsen, J. (2000). *The design of Web usability: The practice of simplicity.* Indianapolis, IN: New Riders Publishing.

Reeves, T.C. & Carter, B.J. (2001). Usability testing and return-on-investment studies: Key evaluation strategies for Web-based training. In B.H. Khan (Ed.), *Web-based training,* (pp. 547-558). Englewood Cliffs, NJ: Educational Technology Publications.

Simich-Dudgeon, C. (1998). Developing a college Web-based course: Lesson learned. *Distance Education, 19*(2), 337-357.

Interface Design Checklist

Page and Site Design

Check if Web pages look good in a variety of Web browsers and in text-based browsers, all recent versions of Internet Explorer and Netscape, and so on.

Browser	Best Viewed By (Type versions)	Not Best Viewed By (Type versions)
Netscape		
Explorer		

Check if the Web documents are available in any of the following formats? (check all that apply):

- ☐ PDF
- ☐ HTML
- ☐ XML
- ☐ Word processed
- ☐ Text file
- ☐ Not applicable
- ☐ Other (specify)

Does the course use the following interface structures? (check all that apply):

- ☐ Text/menu
- ☐ Graphical User Interface (GUI)
- ☐ Voice synthesis and recognition
- ☐ Not applicable
- ☐ Other (specify)

Does the course provide printable transcripts of any streaming audio and video used in the course?

- ☐ Yes
- ☐ No

❏   Not applicable

❏   Other

Do the following elements, if used, complement the textual content of the course? (check all that apply):

❏   Graphics

❏   Audio

❏   Video

❏   Animation

❏   None used

❏   Other (specify)

Do the pages of the course use reasonable blank or white spaces (about 20%) to help readers' eyes move through the content more easily and comfortably? (Note: Insufficient white space can contribute to cluttered screens.)

❏   Yes

❏   No

❏   Not applicable

Is the program attractive and appealing to the eye and ear? (Note: Remember that different people may find different colors or fonts appealing.)

❏   Yes

❏   No

❏   Not applicable

❏   Other

Is the text throughout the course legible?

❏   Yes

❏   No

❏   Not applicable

Throughout the course, are background colors of screens compatible with the foreground colors of the screens (so that they complement rather compete)?

❏   Yes

❑   No
❑   Not applicable

Does the site have a consistent look with the course print materials so the learner can easily make the connection between online course information and correspondence that comes in the mail?
❑   Yes
❑   No
❑   Not applicable

Does the course use a consistent font type across elements such as heading, body text, link, etc.?
❑   Yes
❑   No
❑   Not applicable

Does the course use a standard font type so that text appears the same in different computer platforms and browsers? (e.g., Arial, Times Roman, Helvetica fonts appear the same in different platforms. However, it is a client-side decision; users can display fonts however they want.)
❑   Yes
❑   No
❑   Not applicable

Does the course use a consistent layout including color and the placement of titles and content on Web pages?
❑   Yes
❑   No
❑   Not applicable

Does the choice of graphics enhance the learners understanding of the site's purpose?
❑   Yes
❑   No
❑   Not applicable

How fast do the pages on the course Website load? Do the screens load quickly? Or, must the learner wait for large amounts of graphics, video, audio, and applets to load? (Note: Large images and multimedia files require a long time to download. However, loading speed may vary with users' Internet connection speeds. The course should be designed to use bandwidth efficiently in order to minimize learners' frustration. It is always a good idea to test pages at various Internet connection speeds)

❑   Fast
❑   Fairly fast
❑   Somewhat slow
❑   Very slow

Do parts of the page appear even though the site is not fully loaded?
❑   Yes
❑   No
❑   Not applicable

Do the lessons, assignments and tests take a longer time to complete than the course allow? (Note: It should not be a surprise to anyone that learners from diverse geographical locations with varying Internet connection speeds may take longer to complete learning activities on the Internet than the course designers estimated. To avoid learners' dissatisfaction, course lessons and timed-quizzes or assignments should be tested at dial-up speeds with a representative population.)
❑   Yes
❑   No
❑   Not applicable

Downloading audio and video is often time-consuming. Does the course assign students pre-listening work or other instructional activities while the files are downloading?
❑   Yes
❑   No
❑   Not applicable

Does the site use frames?

❏   Yes

❏   No

❏   Not applicable

If *yes*, does it also have a non-frames version available?

❏   Yes

❏   No

❏   Not applicable

❏   Other (specify)

Does the course give credit to individuals involved in designing and developing the course? (Note: This can be put under a menu item entitled "credit." A credit section is very useful to learners because it allows them to see the credentials of individuals who were involved in the creation of the course. Creditor recognition may not be appropriate for some sites including government and other settings.)

Course Team	Yes	No	NA	Other
Instructor				
Subject Matter Expert (SME) or Content Expert				
Instructional Designer				
Programmer				
Graphic Artist				
Multimedia Developer				
Course Manager				
Other (specify)				

Does the Website provide links to any of the following Websites within the institution? Check all that apply:

❏   Institution's Website

❏   Admissions Office

❏   Financial Aid Office

❏   Academic Departments

❏   Accounting Department

❏   Registrar's Office

❏   Student Services

❑ Student Organizations (Greek, Academic Clubs, etc.)
❑ Information Technology Services
❑ Professional Development
❑ Continuing Education
❑ Other (specify)

Does the course have a link to the instructor's home page and curriculum vitae?
❑ Yes
❑ No
❑ Not applicable

Are colored graphics, if used, clearly interpretable when printed in black and white? (Note: Some users like to print out Web pages to read them later. With a black and white printer, a variety of different colors used in a graphic to distinguish critical parts and functions may not be visible in the print out.)
❑ Yes
❑ No
❑ Not applicable

Does each screen of the course print one printer page?
❑ Yes
❑ No
❑ Not applicable

Content Design

Check if the course uses any of the following ways to gain learner attention? (check all that apply):
❑ Novelty
❑ Animation
❑ Motion (e.g., animated GIFs)
❑ Captioned graphics
❑ Changes in brightness
❑ Contrast between object of interest and its surroundings.

❑   Colors, sounds, and symbols that focus on specific content
❑   Other (specify)

Check if the course uses any of the following ways to improve learner retention? (check all that apply):

❑   Sequenced screens
❑   Meaningfully organized contents
❑   Overviews
❑   Consistent screen layout (consistent placement of title, graphic, textual contents, etc)
❑   Chunked materials, presenting together when appropriate
❑   Introductions and summaries
❑   Other (specify)

Does the course follow the "one idea per paragraph" rule?

❑   Yes
❑   No
❑   Not applicable

Is the text chunked and presented in a way that enables scanning and comprehension? (Note: Throughout the course headings and sub headings should be parallel, short, and logically connected so that readers can scan them.)

❑   Yes
❑   No
❑   Not applicable

Check if any of the following multimedia presentation components are used in the course? (check all that apply):

❑   Text
❑   Graphics
❑   Animation
❑   Audio
❑   Video
❑   Other (specify)

If *yes*, does the mixture of multimedia components contribute to a rich learning environment?

❑   Yes

❑   No

❑   Not applicable

How effectively does the course use multimedia presentation components to create meaningful learning? (check all that apply):

Multimedia Components	Effectiveness			
	High	Moderate	Poor	Other
Text				
Graphics				
Animation				
Audio				
Video				
Other (specify)				

The course content is presented with proper (check all that apply):

❑   Grammar

❑   Punctuation

❑   Spelling

❑   Syntax (how words are put together to form phrases or sentences)

❑   Not applicable

❑   Other (specify)

The course content is presented with appropriate and relevant (check all that apply):

❑   Text

❑   Graphics

❑   Animation

❑   Audio

❑   Video

❑   Other (specify below)

Does the course provide an easy mechanism for electronic publishing for students and instructors?

❑   Yes
❑   No
❑   Not applicable

Navigation

Does the course provide structural aids (i.e., unit, lesson, activities, etc.) to help learners navigate the course?

❑   Yes
❑   No
❑   Not applicable
❑   Other (specify)

Does the course provide a site map (i.e., big picture of the course) to help learners navigate the course?

❑   Yes
❑   No
❑   Not applicable
❑   Other (specify)

To avoid bandwidth bottlenecks, does the course ask students to download large audio, video and graphic files to their hard drives before the instructional events?

❑   Yes
❑   No
❑   Not applicable
❑   Other (specify)

Do pages of the course fit within any graphical browser window without any horizontal or sideways scrolling? (Note: Sideways scrolling can be awkward and annoying at times. It seems to happen when tables are given widths in pixels instead of percentages; a given browser can be too small for the pixels required to display a page, but the percentage is defined as relative to the browser's width.)

❑   Yes
❑   No
❑   Not applicable

Are all links clearly labeled, and do they serve an easily identified purpose, so that learners have enough information to know whether they should click a link?

❑   Yes
❑   No
❑   Not applicable
❑   Other (specify)

Do users have the option to "skip" or "turn off" any animation or media components in the course? (Note: They can be part of the design, but it is a client-side decision too.)

❑   Yes
❑   No
❑   Not applicable
❑   Other (specify)

Does the site contain so many internal links as to be distracting?

❑   Yes
❑   No
❑   Not applicable

Does the site contain so many external links as to be distracting?

❑   Yes
❑   No
❑   Not applicable

Does the site use any icons that are difficult to remember? (Note: In using icons, we should ask "Is it clear what they represent? Does what they represent relate to what they do?")

❑   Yes
❑   No

❏   Not applicable

❏   Other (specify)

Does the course use a consistent color for both unvisited and visited links? (Note: The standard link colors such as 'blue' for unvisited links and 'reddish or purple' for visited links can be used on every page of the course site.)

❏   Yes

❏   No

❏   Not applicable

❏   Other (specify)

Is the course consistent with the use of terminology throughout? (Note: If you use a term or word on one Webpage, it is always wise to use the same term throughout the course.)

❏   Yes

❏   No

❏   Not applicable

❏   Other (specify)

Does the course indicate the size (e.g., 13k, 200k, etc.) of the multimedia files used?

❏   Yes

❏   No

❏   Not applicable

❏   Other (specify)

Does the course have structural flexibility by providing students the choice of multiple pathways through the instruction?

❏   Yes

❏   No

❏   Not applicable

❏   Other (specify)

Does the course offer suggested pathways for the user? (Note: Learners tend to follow links in the course. Therefore, hyperlinking in pages should be well thought out as they suggest pathways for users.)

❑   Yes

❑   No

❑   Not applicable

❑   Other (specify)

How easy is it to navigate the course Website? (Can users move from page to page, and link to link with ease without getting lost or confused?)

❑   Very easy

❑   Fairly easy

❑   Somewhat difficult

❑   Very difficult

Is any part of the course linked to pages that are under construction? (Note: Avoid linking courses to incomplete sites.)

❑   Yes

❑   No

❑   Not applicable

❑   Other (specify)

Are learners informed when they use outside links that lead to different Websites? (Note: In his Distance Educational journal article in 1997, Boshier suggested using signposts or some visual guidance to expedite their return. However, if we open external sites in new browser windows, we do not need any signposts.)

❑   Yes

❑   No

❑   Not applicable

❑   Other (specify)

When a course contains links to sites located in different countries with different cultures (where navigation or expression icons may differ from the learners' native culture), are there any cues on how to adjust to unfamiliar navigation or a different instructional environment? (Boshier, 1997)

❑   Yes
❑   No
❑   Not applicable
❑   Other (specify)

Does the site include a search feature?

❑   Yes
❑   No
❑   Not applicable
❑   Other (specify)

> If *yes*, check all that apply?
> ❑   Internal search feature within course Website
> ❑   External search feature

Does the course use consistent symbols and words as navigation aids? (Boshier, 1997)

❑   Yes
❑   No
❑   Not applicable
❑   Other (specify)

Does the course provide a support mechanism to indicate the progress made? (Note: For example, links that have been visited become a light red color - "bread-crumbing").

❑   Yes
❑   No
❑   Not applicable
❑   Other (specify)

Does the course include features such as context maintenance to automatically return a student to the point where he/she left off during the previous session?

❑   Yes

❑   No

❑   Not applicable

❑   Other (specify)

Does the course provide a progress map or calendar for students to measure their achievement?

❑   Yes

❑   No

❑   Not applicable

❑   Other (specify)

Check all options that apply about menus in the course:

❑   Menus are deep (i.e., more layers)

❑   Menus are shallow

❑   More choices should be available in the menus

❑   Should limit the number of choices in the menus

❑   Not applicable

❑   Other (specify)

Does every page of the course (where frames are not used) have links back to the site's main page?

❑   Yes

❑   No

❑   Not applicable

❑   Other (specify)

Are images used in the course stored on the course Website? (Note: Images saved in places other than the course site may slow down the loading time. Also, if the owner of the image removes it or changes its location, the image will not be found and the "broken image icon" will be displayed (Maddux, 1998).

❑   Yes

❑   No

❑   Not applicable

❑   Other (specify)

Do all the inside links in the course link to the correct locations?

❑   Yes

❑   No

❑   Not applicable

❑   Other (specify)

Do all the outside links in the course link to the correct locations?

❑   Yes

❑   No

❑   Not applicable

❑   Other (specify)

Does the course have any dead links (i.e. inactive links)?

❑   Yes

❑   No

❑   Not applicable

❑   Other (specify)

If *yes*, is there a system or mechanism to check the dead links that may exist within the course and update it on a regular basis?

❑   Yes

❑   No

❑   Not applicable

❑   Other (specify)

If yes, how often are dead links checked?

❑   Daily

❑   Weekly

❑   Monthly

❑   Quarterly

❑   Other (specify)

Does the course have a site that keeps users informed about any changes in URLs used in the course and other course relevant contents? (Note: For this book, a Website at http://BooksToRead.com/elearning/el-update.htm is maintained to inform readers regarding the change of addresses for chapter-related Websites and other corrections.)

❑   Yes

❑   No

❑   Not applicable

Does the course overuse hyperlinks in course pages?

❑   Yes

❑   No

❑   Not applicable

How is the quality of the streaming sound and video used in the course?

Streaming	Quality				
	Excellent	Good	Average	Poor	Very Poor
Sound					
Video					
Other (specify)					

Accessibility

Is the course Website designed to be accessible by a wider user population?

❑   Yes

❑   No

❑   Not applicable

❑   Other (specify)

Are various accessibility barriers considered in the design of the course? (Note: Web pages can be run through Bobby (http://bobby.cast.org) to test Web pages and help expose and repair barriers to accessibility and encourage compliance with existing accessibility guidelines, such as Section 508 and the W3C's WCAG.)

❑   Yes
❑   No
❑   Not applicable
❑   Other (specify)

Is the course Section-508 or W3C compliant?
❑   Yes
❑   No
❑   Not applicable
❑   Other (specify)

If so, at what level?

Does the course use alternate text for the images? (Note: The alternate text for all non-text elements can be read aloud by software for synthesizing speech, and is therefore, essential for visually impaired learners--.)
❑   Yes
❑   No
❑   Not applicable

Does the course provide captions for audio content? (Note: People who cannot hear can read the audio content from the captions--.)
❑   Yes
❑   No
❑   Not applicable

Can various screens of the course be resized to accommodate low-vision users? (Note: Even if the Web pages are designed to a specific screen size, the user can easily resize the screen by using the maximize or minimize option in the browser.)
❑   Yes
❑   No
❑   Not applicable

Are all the colors used in the various screens of the course clearly distinguishable by the visually impaired?

❑   Yes

❑   No

❑   Not applicable

Can users who cannot use the mouse navigate through the e-learning materials using the keyboard instead?

❑   Yes

❑   No

❑   Not applicable

❑   Other (specify)

Does the course use acronyms? (Note: It is always helpful to have the full name represented by the acronym used the first time it appears in the text. For example, UTC (Universal Coordinated Time). Some link acronyms to glossary sites, however it is an extra step for users including individuals with disabilities. E-learning designers should use appropriate scripts in e-learning documents to embed the full name represented by the acronym.)

❑   Yes

❑   No

❑   Not applicable

❑   Other (specify)

If *yes*, check measure(s) taken to solve issues associated with acronyms and accessibility for individuals with disabilities? (check all that apply):

❑   Uses full name represented by the acronym the first time it appears in the text.

❑   Links to glossary

❑   Other (specify)

Usability Testing

Has there been a trial run beforehand with representative users?

❑   Yes

❑   No

❑   Not applicable

❑   Other (specify)

Do users find answers to the most frequently asked questions on the course site within a reasonable amount of time?

❑   Yes

❑   No

❑   Not applicable

Can users easily know where they are and navigate the site without guessing?

❑   Yes

❑   No

❑   Not applicable

Does the course use easy-to-understand terminology?

❑   Yes

❑   No

❑   Not applicable

Can learners easily take a look at or sample each part of the course? (Boshier, 1997)

❑   Yes

❑   No

❑   Not applicable

Is the site designed so that learners can easily get to a specific piece of content (in no more than 3 clicks)?

❑   Yes

❑   No

❑   Not applicable

Chapter 8

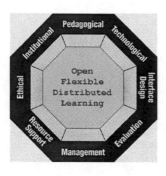

Resource
Support
Issues

Resource support dimension of open, flexible, and distributed learning environment examines the online support and resources required to foster meaningful learning environments. An e-learning institution should have a key component in showing the students that there is an infrastructure that provides support the learners need and gives the security that they are indeed not alone in any of the tasks they are asked to carry out (Fultcher & Lock, 1999). The following in an outline of this chapter:

- Online support
- Resources

Online Support

Both technological and human-based support throughout an e-learning course can help a course maintain momentum and become successful (Hill, 1997).

Online support deals with how an e-learning course can provide online instructional and counseling support and all-purpose technical troubleshooting services. The following in an outline of this section:

- Instructional and counseling support
- Technical support

Instructional and Counseling Support

Individuals who are inexperienced as learners in distance education courses may have a particularly high degree of anxiety at the beginning of the course (Moore & Kearsley, 1996); therefore, e-learning methods can be stressful for some students. Guidance on study skills, self-discipline, responsibility for own learning, time and stress management, and successful degree completion strategies are important components for open, flexible, and distributed learning.

Students should receive guidance on how to organize for online learning. Hart (1999) provides several tips on how to become an effective learner in distance education environments. It would be very useful and beneficial for students to receive some guidance on time and stress management, note taking, reading and writing guides, test anxieties, health and wellness, and so on. Spitzer (2001) notes that he provides an online guidebook about positive distance learning (DL) habits, collected from his DL students in previous courses. Campbell (1999) provides guidance on how to overcome the personal barriers to success in distance learning. Palloff and Pratt (1999) provide guidance on time management.

Institutions should consider time management training for students, instructors, and support staff. Learners should be advised on how they can divide their time into various course-related tasks such as assigned readings, online discussions, individual/group projects, and other assignments. Online courses require self-discipline and self-motivation. Learners should be informed about all possible requirements of online courses so that they can succeed. Appropriate information and timely guidance will reduce the drop-out rate in the online courses and will likely minimize the frustration level and feelings of isolation, which are the two main causes of drop-outs.

Please note that as e-learning becomes more and more used by institutions, the blended learning approach that combines off-line and online forms of learning will also be increasingly used to better supplement online learning. This is good for learners. However, going back and forth between online and off-line learning activities, learners may confuse which resource support services apply to the particular activity. Institutions should clearly inform learners about what support is available online and off-line.

Passmore (2000) noted that e-learning demands a great deal of time and effort from students, faculty, and support staff:

Students often are surprised that distance education often requires more interactivity than classroom-based instruction with which they have more experience. Demands also are high on students' time to maintain the hardware, software, and connectivity for their Web-based courses.

Faculty members meet resistance from students to the increased time demands of distance education that is highly interactive, and they might not pursue an instructional modality that so displeases their customers. Moreover, faculty members often feel that distance education requires too much of their own time to develop and support. They also worry that requirements to accommodate people with disabilities in Web-based instruction will demand even greater time (http://www.aln.org/alnweb/magazine/Vol4_issue2/passmore.htm).

Technical Support

Online technical support is one of the most important support services for e-learning environments. Technical support services must be available to help students log on, upload and download files, for troubleshooting, and so on. It is true that there will always be technical problems due to servers being down, network failure, database crashing, incompatiblity with new versions of software with learning management system (LMS), and so on. Horton (2000) suggests to "plan for disasters" and recommends listing all the disasters that could befall an e-learning course. Technical support should find ways to help learners during a disaster period.

Learners will greatly appreciate getting technical problems solved the easiest and fastest way possible. It is all about doing homework by technical support ahead of time and coming up with meaningful solutions so that learners can easily follow them to fix technical problems and continue their learning process. As we know it is very difficult to be motivated in an open, flexible, and distribute learning environment. Technical problems that learners cannot easily fix or have to wait for long time can put learners behind the schedule, frustrated, and unmotivated. Helping learners during disaster times is the best help.

Resources

The e-learning resources include original documents, public domain books, summaries of or discussions about books in print, reference works (such as foreign language dictionaries), scholarly papers, new concepts, notification of both face-to-face and online conferences, job information, and so on. The following in an outline of this section:

- Online resources
- Off-line resources

Online Resources

Online resources can include multimedia archives, mailing lists and their archives, Frequently Asked Questions (FAQs), glossaries, e-books, dictionaries, calculators, Webliographies, recommended reading lists, digital libraries, computer tutorials, online experts, journals, magazines, newsletters, newspapers, documents, Webfolios (i.e., an electronic version of the portfolio), personal journals (i.e., Web logs or blogs), knowledge management, and so on.

All online resources should be limited to what learners need for specific tasks in the course. Learners should be provided with some guidelines on how to assess the quality and utility of information available online. French (1999) noted that authority, accuracy, timeliness, and bias are four quality indicators for resources on the Web. All resources should be relevant and essential. An overwhelming volume of information and irrelevant resources may frustrate the learners. Students should be knowledgeable about the sources of resources. They should know from the extension of the Website address whether or not particular information is provided by an organization (.org), company (.com), government, military, or educational institution.

Various online resources with some examples are provided below. Institutions should consider either creating these resources (if possible) or linking them whenever appropriate.

Offline Resources

Offline resources can include books, journals, magazines, newsletters, newspapers, documents, reference works, experts, and so on. Institutions offering e-

Table 1.

Online Resources	Examples
Multimedia archives	Multimedia Educational Resource for Learning and Online Teaching (MERLOT) is a free and open resource designed primarily for faculty and students of higher education. Links to online learning materials are collected along with annotations such as peer reviews and assignments. URL: http://www.merlot.org
Mailing lists and their archives	Anyone can subscribe (generally at no charge) to an e-mail mailing list on a particular subject or subjects and to post messages. The Distance Education Online Symposium Listserv (DEOS-L) is a moderated listserv that facilitates discussion of current issues in distance education. URL: http://www.ed.psu.edu/acsde/deos/deos-l/deosl.asp
Frequently Asked Questions (FAQs)	Hillsborough Community College in Florida hosts an e-learning FAQs site for students. URL: http://www.hcc.cc.fl.us/dislearn/Summer_2003/studfaqs.htm
Glossaries	PlasmaLink Web Services provides the Glossary of Instructional Strategies as a resource for all educators. URL: http://glossary.plasmalink.com/glossary.html
e-books	University of Virginia's E-Book Library has hundreds of publicly-available e-books for classic British and American fiction, major authors, children's literature, American history, Shakespeare, African-American documents, the Bible, and much more. URL: http://etext.lib.virginia.edu/ebooks/ebooklist.html
Dictionaries	The Resources for the Study of Norwegian site provides online dictionaries in Norwegian and other languages. URL: http://employees.csbsju.edu/tnichol/norwegian.html#dict
Calculator	The State University of New York developed an online calculator for prospective students to determine how much money they can save by taking a college course from their home. URL: http://sln.suny.edu/sln/public/original.nsf/a6b56cc3058e682485256c790066b2d5?OpenForm
Webliographies	The Maritime History Webliography at the University of North Carolina attempts to organize and classify those resources currently available on the Internet with some connection to maritime history. URL: http://www.ils.unc.edu/maritime/mhiweb/webhome.shtml
Recommended reading lists	Recommended reading list in educational technology, instructional technology, training, distance education, online learning, online education, open learning design, k-12, multimedia, and user interface design. Also, it contains a top ten list by experts in the field. URL: http://BooksToRead.com/et.htm
Digital libraries	Institutions with online programs should consider either establishing their own digital libraries, partnership (consortia) digital libraries, or service agreement with other digital libraries to provide digital library support to their online students. These digital libraries should be created to insure that all available resources are complete, highly searchable and richly formatted (Schmitz, 2001). The California Digital Library (CDL) is an additional "co-library" of the University of California (UC) campuses, with a focus on digital materials and services. CDL is a collaborative effort of the 10 UC campuses. URL: http://www.cdlib.org/
Computer tutorials	The Department of Sociology at University of California (Davis) developed a computer tutorial titled "Add a Network Printer." URL: http://sociology.ucdavis.edu/tutorials/Add_a_Network_Printer_viewlet_swf.html

Table 1. (continued)

Experts online	Experts Online hosted by Local Initiatives Support Corp is an interactive forum for professional discussion among industry experts, as well as national and local practitioners. The Experts Online live event (free of charge) is a support and training service to community development practitioners nationwide. URL: http://www.liscnet.org/resources/experts_index.shtml
Journals and Magazines	*The Technology Source* (ISSN 1532-0030), a peer-reviewed bimonthly periodical published by the Michigan Virtual University, is to provide thoughtful, illuminating articles that will assist educators as they face the challenge of integrating information technology tools into teaching and into managing educational organizations. URL: http://ts.mivu.org/
Newsletters	The "LD OnLine Newsletter" provides up-to-date information for the field of learning disabilities. UR L: http://www.ldonline.org/subscribe.html
Newspapers	The Internet Public Library has links to online newspapers from around the world. URL: http://www.ipl.org/div/news/
Personal journals (i.e., Web logs or blogs)	Blogs or Weblogs are an informal personal Web sites which can be used as powerful e-learning resources. Weblogs are sometimes called Web journals. An increasing number of people are blogging every day. Therefore, there are numerous Weblogs in various topics which are updated regularly. Professor Ray Schroeder at the University of Illinois at Springfield scans the news daily for items of relevance in the field of new communication technologies, educational technologies, and online learning. He created three Weblogs (Online Learning Update, Education Technology, and Techno-News) to keep his technologies seminars and classes up-to-date and fresh for his students. URL: http://people.uis.edu/rschr1/bloggerinfo.html
Knowledge management	The Web site "Teaching and Implementing Knowledge Management Programs" provides many examples of companies and organizations that are implementing knowledge management. URL: http://www.icasit.org/kmclass/teaching/

learning to geographically disperse remote learners should provide suggestions or information about where to find library resources since many cannot use the host institution's library due to distance. Also, the host institution should consider joining a consortium of libraries worldwide so their students can visit and loan books.

Question to Consider

Can you think of any specific resource support issues not covered in this chapter?

Activity

1. Using Internet search engines, locate at least one article relevant to in any one of the following online resources, and analyze the article from the perspective of its usefulness in e-learning:

    *   Multimedia archives
    *   Mailing lists and their archives
    *   Frequently Asked Questions (FAQs)
    *   Glossaries
    *   E-books
    *   Dictionaries
    *   Calculators
    *   Webliographies
    *   Recommended reading lists
    *   Digital libraries
    *   Computer tutorials
    *   Experts online
    *   Journals and Magazines
    *   Newsletters
    *   Newspapers
    *   Personal journals (i.e., Web logs or blogs)
    *   Knowledge management

2. Locate an online program and review its facilities for online support and resources using the relevant resource support checklist items at the end of Chapter 8.

References

Campbell, S.M. (1999). Understanding your needs: Overcoming the personal barriers to success in distance learning. In G.P. Connick (Ed.), *The distance learner's guide.* Upper Saddler River, NJ: Prentice Hall.

French, D. (1999). Skills for developing, utilizing and evaluating Internet-based learning. In D. French, C. Hale, C. Johnson, & G. Farr (Eds.), *Internet based learning: An introduction and framework for higher education and business,* (pp. 63-86). Sterling, VA: Stylus Publishing LLC.

Fulcher, G. & Lock, D. (1999). Distance education: The future of library and information services requirements. *Distance Education: An International Journal, 20*(2), 313-329.

Hart, J. (1999). Improvng distance learning performance. In G.P. Connick (Ed.), *The distance learner's guide.* Upper Saddler River, NJ: Prentice Hall.

Hill, J.R. (1997). Distance learning environments via the World Wide Web. In B.H. Khan (Ed.), *Web-based instruction.* Englewood Cliffs, NJ: Educational Technology Publications.

Horton, W.K. (2000). *Designing Web-based training: How to teach anyone anything, anywhere, anytime.* John Wiley & Sons.

Moore, M. & Kearsley, G. (1996). *Distance education: A systems view.* Belmont, CA: Wadsworth.

Palloff, R.M. & Pratt, K. (1999). *Building learning communities in cyberspace.* San Francisco, CA: Jossey-Bass Publications.

Passmore, D.L. (2000). Impediments to adoption of Web-based course delivery among university faculty. The Pennsylvania State University. Retrieved January 24, 2003, from *http://www.aln.org/alnweb/magazine/Vol4_issue2/passmore.htm*

Schmitz, J. (2001). Needed: Digital libraries for Web-based training. In B.H. Khan (Ed.), *Web-based training,* (pp. 391-394). Englewood Cliffs, NJ: Educational Technology Publications.

Spitzer, D.R. (2001). Don't forget the high-touch with the high tech in distance learning. *Educational Technology, 41*(2), 51-55.

Resource Support Checklist

Online Support

Check if the course provides any of the following informational and communication options for any of the following individual(s)/office(s). (check all that apply):

Role of Individual	Method of Communication				
	Email	Live Chat	Telephone		Other
			Regular*	Toll free**	
Course Coordinator					
Instructor					
Tutor					
Graduate Assistant					
Discussion Facilitator/Moderator					
Copyright Coordinator					
Guest speaker (or outside expert)					
Counselor					
Career counseling services					
Technical support					
Learning resources					
Tutoring service					
Library services					
Admission office					
Registration service					
Bursar office					
Financial aid					
Bookstore					
Other					

*   *Should always include the area code with the telephone number*

**   *If an institution decides to provide toll free phone call services for learners, then it must be available to all learners regardless of their diverse locations. For example, a US institution offering e-learning courses to learners worldwide should provide alternatives to toll free phone call such as 1-800, 1-877 or 1-866 to learners outside of the USA and Canada. Please note that 1-800 or 1-877 does not work outside the USA and Canada. Also, TTY phone services should be available for the deaf or the hearing impaired.*

Does the course include a Frequently Asked Question (FAQ) page?

❑ Yes

❑ No

❑ Not applicable

Does the institution conduct a pre-assessment survey to identify if learners have the necessary skills for online learning?

❑ Yes

❑ No

❑ Not applicable

Is there an *introduction to online* or a similar course offered by the institution that covers any of the following skills? Check all that apply:

Skills	Yes	No	NA	Comment
Study skills (reading and writing guides, note taking, etc.)				
Self-discipline				
Time management				
Stress management				
Health and wellness				
Test anxieties				
How to use available resources				
Other				

Check if the course provides counseling sessions for distance learners by any of the following informational and communication options. (check all that apply):

Counseling Issues	Method of Communication								E-Mail	Other
	Face-to-Face Meeting (Hours Per Week)				Telephone (Hours Per Week)					
	1-3	4-6	7-10	Other	1-3	4-6	7-10	Other		
Guidance on study skills										
Time management										
Stress management and personal problems										
Career guidance										
Other (specify)										

Are students required to submit a Counseling Appointment Request Form (CARP) to the counselor prior to a counseling session? (Note: At the Athabasca University in Canada, students are required to complete and submit a CARP to request a session with counselor. Students are informed of the confirmation of appointments via e-mail within 7 business days.)

❑   Yes

❑   No

❑   Not applicable

Does the course provide any guidance to students on how to organize for online learning?

❑   Yes

❑   No

❑   Not applicable

Does the instructor assist students who encounter problems in completing their assignments?

❑   Yes

❑   No

❑   Not applicable

Do students receive guidance on any of the following skill(s)?

❑   Yes

❑   No

❑   Not applicable

If *yes*, check all that apply:

❑   Ability to work alone

❑   Ability to learn without face to face classroom interaction

❑   Ability to do collaborative work with never-met individuals

❑   Not applicable

❑   Other

Does the instructor/staff contact students (who fail to participate in regular online learning activities for the course) to see if they are encountering problems?

❏   Yes

❏   No

❏   Not applicable

   If *yes*, how are students contacted? Check all that apply:
   ❏   E-mail
   ❏   Phone
   ❏   Fax
   ❏   Letter
   ❏   Other

Does the course provide any information or ideas about how many hours (approximately) per week students are expected to spend on course assignments?

❏   Yes

❏   No

❏   Not applicable

Do students receive any guidance on how to search course relevant resources on the Web using search engines?

❏   Yes

❏   No

❏   Not applicable

Do students receive any guidance on the quality and reliability of online resources they find using search engines?

❏   Yes

❏   No

❏   Not applicable

Does the course provide someone other than the instructor who can assist with student problems regarding learning tasks? (Note: In addition to the instructional team, the course can use peer to peer groups for such situations.)

❏   Yes

❏   No

❏   Not applicable

If *yes*, please specify

Does the instructor provide timely responses to student queries?

❏   Yes

❏   No

❏   Not applicable

Does the course provide someone other than the instructor who can help students with problems?

❏   Yes

❏   No

❏   Not applicable

Does the institution regularly review the effectiveness of counseling services?

❏   Yes

❏   No

❏   Not applicable

❏   Other (specify)

If *yes*, check if any of the following used for collecting data. (check all the apply):

❏   Student surveys

❏   E-mail communications

❏   Telephone

❏   Other (specify)

Does the course provide links to Websites that provide subject-related job postings? (Note: Jones International University provides students with links to Websites where they can either post their resume online or review job postings from companies around the country (http://www.e-globallibrary.com/eprise/ main/egloballibrary/demo/index).

☐   Yes

☐   No

☐   Not applicable

☐   Other (specify)

Check if the course provides technical support for distance learners by any of the following informational and communication options. (check all that apply):

Days	Method of Communication									
	Telephone (Hours)				Live Chat (Hours)				E-Mail	Other
	1-3	*4-6*	*7-10*	*Other*	*1-3*	*4-6*	*7-10*	*Other*		
Monday										
Tuesday										
Wednesday										
Thursday										
Friday										
Saturday										
Sunday										

Does the course provide troubleshooting (or expert technical support from specialized staff) assistance or a help line? (Note: If the course is hosted on a vendor's LMS (learning management system), then it needs to be very clearly described to learners about who provides the technical support for the LMS.)

☐   Yes

☐   No

☐   Not applicable

Does the course provide round-the-clock (24/7) technical support?

☐   Yes

☐   No

☐   Not applicable

If the course is offered in multiple languages, are the round-the-clock (24/7) technical support services available in all these languages?

❑   Yes

❑   No

❑   Not applicable

Does the course provide clear guidelines to the learners on what support can and cannot be expected from a help line? (Note: For example, things the student is responsible for, and things that the student can expect the help line to solve.)

❑   Yes

❑   No

❑   Not applicable

Do students receive any guidance on how to set up hardware equipment for desktop video conferencing (if needed for the course)?

❑   Yes

❑   No

❑   Not applicable

Does the course provide a print-based User Guide for learners?

❑   Yes

❑   No

❑   Not applicable

Do students receive any guidance on how to do the following?

❑   Send and respond to e-mail

❑   Send e-mail attachments

❑   Open files in e-mail

❑   Install required software

❑   Scan a picture

❑   Print within Webpage frames

❑   Create an online presentation using presentation software

❑   Transfer and receive files between the learner's desktop and the institution's server

❑   Organize bookmarks in the browser

❑   Other (specify)

Does the course provide technical support materials on the Web?

❑   Yes

❑   No

❑   Not applicable

Does technical support send e-mail answers to a question from one student about a general technical issue to all students?

❑   Yes

❑   No

❑   Not applicable

Check if any of the following forms of technical support are available for students (check all that apply):

❑   Help-desk technician on duty

❑   Interactive training video (online)

❑   Interactive training video (mailed to learners)

❑   Call-in lines

❑   Assistance from instructor/tutors

❑   Other

❑   Not applicable

If asynchronous help is provided, then how soon can learners expect to get answers to their e-mail, phone message, or fax inquiries from the Technical Support Staff?

❑   Within 6 hours

❑   Within 12 hours

❑   Within 24 hours

❑   Within 36 hours

❑   Within 48 hours

❑   Does not respond to e-mail messages

❑   Does not return phone messages

❑   Does not respond to fax messages
❑   Not applicable
❑   Other (specify)

In case of technical difficulties, what other ways can students submit their assignments?
❑   Fax
❑   Mail
❑   E-Mail attachment
❑   Not applicable
❑   Other (specify below)

Online Resources

Online Resources	Examples
Multimedia archives	Multimedia Educational Resource for Learning and Online Teaching (MERLOT) is a free and open resource designed primarily for faculty and students of higher education. Links to online learning materials are collected along with annotations such as peer reviews and assignments. URL: http://www.merlot.org
Mailing lists and their archives	Anyone can subscribe (generally at no charge) to an e-mail mailing list on a particular subject or subjects and to post messages. The Distance Education Online Symposium Listserv (DEOS-L) is a moderated listserv that facilitates discussion of current issues in distance education. URL: http://www.ed.psu.edu/acsde/deos/deos-l/deosl.asp
Frequently Asked Questions (FAQs)	Hillsborough Community College in Florida hosts an e-learning FAQs site for students. URL: http://www.hcc.cc.fl.us/dislearn/Summer_2003/studfaqs.htm
Glossaries	PlasmaLink Web Services provides the Glossary of Instructional Strategies as a resource for all educators. URL: http://glossary.plasmalink.com/glossary.html
e-books	The University of Virginia's E-Book Library has hundreds of publicly-available ebooks for classic British and American fiction, major authors, children's literature, American history, Shakespeare, African-American documents, the Bible, and much more. URL: http://etext.lib.virginia.edu/ebooks/ebooklist.html
Dictionaries	The Resources for the Study of Norwegian site provides online dictionaries in Norwegian and other languages. URL: http://employees.csbsju.edu/tnichol/norwegian.html#dict
Calculator	State University of New York developed an Online Calculator for prospective students to determine how much money they can save by taking a college-course from their home. URL: http://sln.suny.edu/sln/public/original.nsf/a6b56cc3058e682485256c790066b2d5?OpenForm
Webliographies	The Maritime History Webliography at the University of North Carolina attempts to organize and classify those resources currently available on the internet with some connection to maritime history. URL: http://www.ils.unc.edu/maritime/mhiweb/webhome.shtml
Recommended reading lists	A recommended reading list for educational technology, instructional technology, training, distance education, online learning, online education, open learning design, k-12, multimedia and user interface design. Also, contains Top Ten List by experts in the field. URL: http://BooksToRead.com/et.htm
Digital libraries	Institutions with online programs should consider either establishing their own digital libraries, partnership (consortia) digital libraries, or service agreement with other digital libraries to provide digital library support to their online students. These digital libraries should be created to insure that all available resources are complete, highly searchable and richly formatted (Schmitz, 2001). The California Digital Library (CDL) is an additional "co-library" of the University of California (UC) campuses, with a focus on digital materials and services. CDL is a collaborative effort of the ten UC campuses. URL: http://www.cdlib.org/

(continued from previous page)

Computer tutorials	The Department of Sociology at University of California (Davis) developed a computer tutorial entitled "<u>Add a Network Printer</u>." URL: http://sociology.ucdavis.edu/tutorials/Add_a_Network_Printer_viewlet_swf.html
Experts online	Experts Online hosted by the Local Initiatives Support Corporation is an interactive forum for professional discussion among industry experts, as well as national and local practitioners. Experts Online live event (free of charge) as a support and training service to community development practitioners nationwide. URL: <u>http://www.liscnet.org/resources/experts_index.shtml</u>
Journals and Magazines	*The Technology Source* (ISSN 1532-0030), a peer-reviewed bimonthly periodical published by the Michigan Virtual University, provides thoughtful, illuminating articles that will assist educators as they face the challenge of integrating information technology tools into teaching and into managing educational organizations. URL: http://ts.mivu.org/
Newsletters	The *LD OnLine Newsletter* provides up-to-date information for the field of learning disabilities. UR L: http://www.ldonline.org/subscribe.html
Newspapers	The Internet Public Library has links to online newspapers from around the world. URL: http://www.ipl.org/div/news/
Personal journals (i.e., Web logs or blogs)	Blogs or Weblogs an informal personal Websites which can be used as powerful e-learning resources. Weblogs are sometimes called Web journals. An increasing number of people are blogging every day. Therefore, there are numerous Weblogs in various topics which are updated regularly. Professor Ray Schroeder at the University of Illinois at Springfield scans the news daily for items of relevance in the field of new communication technologies, educational technologies, and online learning. He created three Weblogs (Online Learning Update, Education Technology, and Techno-News) to keep his technologies seminars and classes up-to-date and fresh for his students. URL: <u>http://people.uis.edu/rschr1/bloggerinfo.html</u>
Knowledge management	The Website entitled "Teaching and Implementing Knowledge Management Programs" provides many examples of companies and organizations that are implementing knowledge management. URL: http://www.icasit.org/kmclass/teaching/

Does the course have (or links to) any of the following online resources? (check all that apply):

- ❑   Multimedia archives
- ❑   Mailing lists and their archives
- ❑   Newsgroups
- ❑   FAQs
- ❑   Glossaries
- ❑   E-Books
- ❑   Dictionaries
- ❑   Calculators

- ❑ Webliographies
- ❑ Recommended reading lists
- ❑ Databases
- ❑ Digital libraries
- ❑ Computer tutorials
- ❑ Experts online
- ❑ Electronic books
- ❑ Journals
- ❑ Magazines
- ❑ Newsletters
- ❑ Newspapers
- ❑ Documents
- ❑ Personal journals (i.e., web logs or blogs)
- ❑ Knowledge management

Are all available resources organized logically with course relevant categories?

- ❑ Yes
- ❑ No
- ❑ Not applicable

Do all categories of available online resources have brief descriptions about their contents?

- ❑ Yes
- ❑ No
- ❑ Not applicable

Does the course have a searchable *course glossary*?

- ❑ Yes
- ❑ No
- ❑ Not applicable

Does the course glossary include acronyms?

- ❑ Yes

❑ No
❑ Not applicable

Does the course provide an online bookstore (i.e. a means of purchasing course books online)?
❑ Yes
❑ No
❑ Not applicable
❑ Other (specify below)

Does the course provide access to other sources of information related to the course content (e.g. video, audio materials)?
❑ Yes
❑ No
❑ Not applicable

If *yes*, please describe:

Does the course provide summaries and reviews of online discussions?
❑ Yes
❑ No
❑ Not applicable

Does the course provide links to a variety of search engines?
❑ Yes
❑ No
❑ Not applicable

Does the course have any examples of course-related professional work available at other Websites?
❑ Yes
❑ No
❑ Not applicable

Are the external links to resources appropriately related to the context of the content?

❑   Yes

❑   No

❑   Not applicable

Do the external links to resources increase the credibility of the course?

❑   Yes

❑   No

❑   Not applicable

Are the external links checked on a regular basis to make sure they still work?

❑   Yes

❑   No

❑   Not applicable

Does the course provide a student annotation facility for students to make notes for future reference?

❑   Yes

❑   No

❑   Not applicable

Does the course make use of EPSS - Electronic Performance Support System (software designed to improve productivity by providing immediate on the job access to learning and information) as a research tool?

❑   Yes

❑   No

❑   Not applicable

Does the course provide examples of previous students' work on the Web?

❑   Yes

❑   No

❑   Not applicable

If *yes*, select all that apply and circle whether searchable and browsable:
- ❑ Projects (searchable browsable)
- ❑ Papers (searchable browsable)
- ❑ Text dialogue from discussion forums (searchable browsable)
- ❑ Text dialogue from online conferencing exchanges (searchable browsable)
- ❑ Other

Does the institution's library have library resources online?
- ❑ Yes
- ❑ No
- ❑ Not applicable

If *yes*, do students have access to its databases via the Internet or other network?
- ❑ Yes
- ❑ No
- ❑ Not applicable

Does the institution have an online means of borrowing books and other resources?
- ❑ Yes
- ❑ No
- ❑ Not applicable

If *yes*, describe the following:
- ❑ How long books or other resources can be kept
- ❑ How books or other resources can be returned (e.g., via regular mail or other means)

Does the institution have a digital library of its own?
- ❑ Yes
- ❑ No

❏   Not applicable

If *no*, does the institution have a partnership with other institutions to use their digital libraries?

❏   Yes

❏   No

❏   Not applicable

❏   Other (specify)

Do the online resource Websites provide bibliographies or lists of references to indicate the original sources of materials included in their sites?

❏   Yes

❏   No

❏   Not applicable

❏   Other (specify)

Check if students have access to any of the following: (Check all that apply):

❏   Online catalogs

❏   Periodical indexes

❏   Bibliographic databases

❏   Other

Check if the institution provides a special librarian who is available to assist learners at a distance using any of the following informational and communication options. (Note: AskUsNow is a live online interactive library service provided through a partnership of Maryland public, academic, and special libraries. Students can ask questions to librarians about their research in real time via the Internet—24 hours a day, seven days a week. URL: http://www.askusnow.info/) (check all that apply):

Days	Method of Communication									
	Telephone (Hours)				Live Chat (Hours)				E-Mail	Other
	1-3	*4-6*	*7-10*	*Other*	*1-3*	*4-6*	*7-10*	*Other*		
Monday										
Tuesday										
Wednesday										
Thursday										
Friday										
Saturday										
Sunday										

Does the institution make special arrangements with local libraries for distance learners to have access to library resources?

❑   Yes

❑   No

❑   Not applicable

Do remote students receive special training on how to access library resources electronically (e.g., library orientation)?

❑   Yes

❑   No

❑   Not applicable

Do bandwidth limitations affect remote access to library resources?

❑   Yes

❑   No

❑   Not applicable

Does the course provide an on line course reference manual for site mechanics (how to use the site)?

❑   Yes

❑   No

❑   Not applicable

If needed for the coursework, does the course provide on line tools such as a calculator for students to use?

❑   Yes

❑   No

❑   Not applicable

Does the institution have a knowledge management (KM) site?

❑   Yes

❑   No

❑   Not applicable

Does the course use Weblogs as resources for learning?

❏   Yes

❏   No

❏   Not applicable

Offline Resources

Does the course require any of the following off-line resources?

❏   Yes

❏   No

❏   Not applicable

If *yes*, check all that apply:

❏   Dictionaries

❏   Glossaries

❏   Books

❏   E-Books

❏   Papers

❏   Maps

❏   .pdf files which can be downloaded for later reading

❏   Other (specify below)

Does the course require off-line reading assignments?

❏   Yes

❏   No

❏   Not applicable

If *yes*, how is the reading material accessed?

Does the host institution's library have a system of getting books and other materials for distance students via interlibrary loan?

❑   Yes

❑   No

❑   Not applicable

Does the institution provide information on how to get region-wide borrowers cards to borrow books from other academic libraries?

❑   Yes

❑   No

❑   Not applicable

Does the library fax documents to students?

❑   Yes

❑   No

❑   Not applicable

Chapter 9

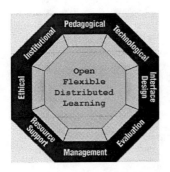

Evaluation Issues

Evaluation in e-learning should focus on the people, processes, and products of e-learning. Evaluation issues of e-learning should consider how e-learning and blended learning materials are planned, designed, developed, delivered, and maintained; how well courses are taught and supported; how well program and institutional level services are provided; how e-learning programs are viewed by stakeholders and how well learners learned the materials. To explore theses issues both formative and summative evaluation strategies can be used:

- Evaluation of e-learning content development process
- Evaluation of e-learning environment
- Evaluation of e-learning at the program and institutional levels
- Assessment of learners

Evaluation of E-Learning Content Development Process

As discussed in Chapter 3, the e-learning content development process includes the planning, design, production, and evaluation of e-learning contents (see Figure 1).

The people, processes and products involved in e-learning content process should be thoroughly evaluated (see Figure 2).

Figure 1. Content development process

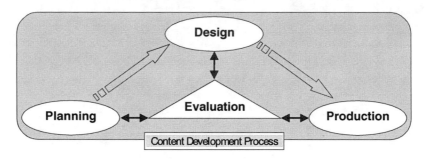

Figure 2. People process product continuum for content development process

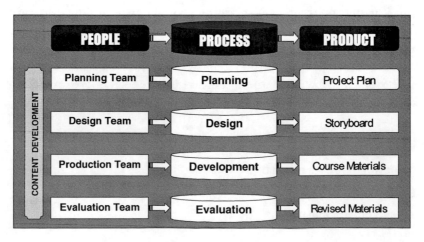

People

Evaluation of individuals involved in the planning, design, production, and evaluation stages of the content development process should be conducted. These individuals may include (but are not limited to): program director, project manager, business developer, consultant/advisor, research and design coordinator, content or subject matter expert, instructional designer, interface designer, copyright coordinator, evaluation specialist, production coordinator, course integrator, programmer, editor, graphic artist, multimedia developer, photographer/videographer (cameraman), learning objects specialist, quality assurance, pilot subjects, and so on.

Process

Evaluation of the planning, design, production, and evaluation stages of the content development process is critical. The performance level of the various types of the evaluation process, including; content review, rapid prototype, alpha class and beta class should be reviewed, as should the performance level of the various tools and services used during the content development process. These tools and services may include: content development /authoring tool, learning management system, screen reader software, accessibility evaluation tool, network server, hardware vendor services, and software vendor services.

Product

In this section, evaluation of the products of the planning, design, production, and evaluation stages of the content development process is conducted. For example, the project plan, storyboard, course materials, and revised course materials are products of the planning, design, production, and evaluation stages of the content development process respectively.

Evaluation of E-Learning Environment

As discussed in Chapter 3, e-learning environment includes e-learning delivery and maintenance process (see Figure 3). The people, processes, and products involved in delivering, maintaining, teaching, and supporting e-learning courses should be thoroughly evaluated (see Figure 4).

Figure 3. Delivery and maintenance of e-learning environment (highlighted)

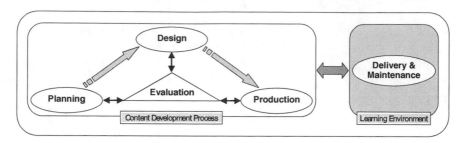

Figure 4. People process product continuum for the delivery and maintenance process

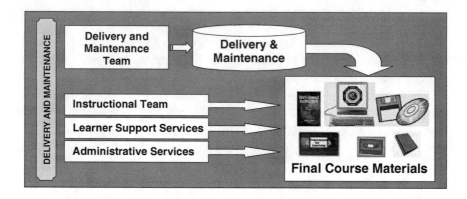

Evaluation of an e-learning environment is different than the evaluation of a traditional face to face classroom based environment. In traditional face to face classes, learners usually evaluate their instructors — which makes sense, since instructors are the ones who taught these courses with their own lecture materials. However, e-learning at an open, flexible, and distributed e-learning environment is a different paradigm. E-learning courses are usually designed, developed, delivered, and supported by various individuals and support units within an institution. Instructors are just one part of the e-learning environment.

Therefore, just evaluating the instructor will not provide the total picture of the e-learning environment as the instructor's performance is systemically dependent on the quality of course design, support services and efficiency of technology infrastructure. Like the proverb, "it takes a whole village to raise child," the learning at a distance is fostered by the instructor and other support staff including, tutor, technical support person, librarian, counselor, and registration staff.

Several e-learning evaluation instruments encompassing the various dimensions of e-learning environment are in the process of being developed. Information about these instruments is available at: http://BooksToRead.com/elearning/evaluation.

Arizona State University Online (ASUonline) put together a sample online course evaluation form. Areas covered in the evaluation form include, institutional support, course development, teaching and learning, course structure, student support, value, flexibility, and convenience (http://asuonline.asu.edu/courseevaluation/sample.cfm).

Evaluation of instruction and learning environment for e-learning should focus on how its customers (i.e., learners) and the community it serves (i.e., market) feel about the overall performance of its e-learning offerings. Students' feedback provides an accurate portrayal of an e-learning environment. Learners are only knowledgeable about the delivery stage of the course and not familiar with the course design process. Therefore, institutions should develop appropriate evaluation methods to get learners' feedback on instructional and support services and delivery of e-learning. Whenever appropriate, institution should also get learners' feedback on the design of the e-learning environment.

The following is an outline of this section:

1. Evaluation of e-learning delivery and maintenance
2. Evaluation of course offerings

Evaluation of E-Learning Delivery and Maintenance

The design of the course greatly influences the roles technical and support services play in e-learning environments. This section covers the evaluation of how well e-learning environments are delivered and maintained (Also discussed in *E-Learning Delivery and Maintenance* section of Chapter 3).

Course Offerings

As discussed in Chapter 3, the course offerings include all e-learning and blended learning materials available to learners. In this section, evaluation of how course materials are provided, taught and supported is discussed. Figure 5 graphically represents e-learning environment at the eTQM College (http://etqm.net).

Figure 5. E-learning environment

Evaluation of e-learning environment includes: (1) performances of instructional team (e.g., instructor or trainer), instructor assistant, tutor, discussion facilitator/moderator, learning objects specialist, copyright coordinator, guest speaker (or outside expert); (2) learner support services staff including systems administrator, server/database programmer, customer service, technical support specialist, library services, counseling services, and so on, and (3) administrative services including admissions, registration, payment, bookstore, financial aid, and so on.

Evaluation of Instructional Team

Evaluation of instructor and other members of the instructional team (if any) can be done by asking learners to complete an instructor evaluation form consisting of both open-ended and multiple choice items. In addition to learners' feedback, an instructor's performance can also be evaluated by his/her participation in both asynchronous (e.g., discussion forums postings) and synchronous (e.g., chat room transcripts) activities. The URLs of instructor evaluation for online courses are listed below:

Institution Name	URL
Ohio University Online Course Evaluation	http://www.ohiou.edu/ouonline/evaluation.html
Montana State University-Billing Online	http://www.msubillings.edu/support101/eCollege/courseevaluation.htm

Evaluation of Learner Support Services

Evaluation of learner support including technical support specialist, library services, counseling services, and customer service can be accomplished by asking learners to complete evaluation form consisting of both open-ended and multiple choice items. Arizona State University Online (ASUonline) provides a sample online course evaluation form which has items on student support (http://asuonline.asu.edu/courseevaluation/sample.cfm).

Evaluation of Administrative Support

Evaluation of administrative support (e.g., admission, registration, payment, bookstore, financial aid, etc.) can be accomplished by asking learners to complete an evaluation form consisting of both open-ended and multiple choice items.

Evaluation of E-Learning at the Program and Institutional Levels

In an institution, all professional development, certificate, and degree programs enjoy some types of centralized services. For example, students can register via the institution's registrar office. Depending on the institution, some services may be offered at the institution or at the program level. Advising and orientation can be provided by individual programs. Some services (e.g., marketing) can either be provided by individual programs or by the institution. To students, it does not matter who is providing these services. What matters to them is how well they are provided. There is tremendous need for a comprehensive evaluation of e-learning, which should include the performance of each individual and support unit involved in providing various e-learning services.

Institutions should develop evaluation criteria on all aspects of e-learning including: course development and delivery, learning environment, and support services. For example, institutions can develop evaluation forms by following accreditation standards programs set by regional and international accrediting agencies. Western Cooperative for Educational Telecommunications (WCET) developed "Best Practices for Electronically Offered Degree and Certificate Programs" which might be of interest to institutions offering e-learning: http://www.ncahigherlearningcommission.org/resources/electronic_degrees/Best_Pract_DEd.pdf

Based on the analysis of learners' feedback, an institution has a better understanding about the status of its e-learning offerings. Montana State University at Billings uses a Student Support Services survey to receive students' feedback on the overall program including: program information services, program administration services, program orientation services, program participation services, and program evaluation services (http://www.msubillings.edu/support101/eCollege/supportsurvey.htm).

Assessment of Learners

Assessment pertains to authenticity, reliability, formats (e.g., multiple choice, essays, case studies, electronic portfolios, etc.), and test characteristics (e.g., adaptive and randomized). A variety of evaluation and assessment tools can be incorporated into an e-learning course. Individual testing, participation in group discussions, questions, and portfolio development can all be used to evaluate students' progress. Assessment in e-learning should be congruent with the pedagogical approach of the course.

Considering the open and flexible nature of the e-learning environment, assessment of learners at a distance can be a challenge. Issues of cheating are a major concern (Wheeler, 1999). Questions such as: "Are students actually doing the work?" (Hudspeth, 1997) and "How do we know we are assessing fairly and accurately?" (Wheeler, 1999) will always be of concern for online-learning environments. Assessing learners from their participation in online discussion can be very difficult especially when some students "lurk." Romiszowski and Chang (2001) notes that a lurker may be benefiting just as much as the silent students in class who learn from the comments and questions of other students.

It is important to clearly indicate assignment due dates for learners from geographically diverse time-zones. The following observation by a distance education faculty member is very interesting when it comes to due dates:

My experience has been that students are more successful if they are given due dates for assignments. The course can remain self-paced to a certain extent in that students are certainly welcome to do assignments earlier than the due date.

We found that when students were not given interim due dates that the majority of them procrastinated. When faced with the decision of completing an assignment for another class that had a due date or completing an assignment for the class without a due date until the end of the course, the students chose the former to spend their time. Consequently, about three weeks before the end of the term, students belatedly realized they could not possibly complete all the assignments and would drop the class.

Unfortunately, I did not do a formal study of this phenomenon, so I am only reporting my observations over a two-year period. I can say that when the professors went from no due dates to interim due dates they all reported experiencing the same thing. When there were no due dates, I observed that class attrition accelerated in the last third of the term. Once interim due dates were in place, attrition occurred in the first half of the term as in all other classes. Needless to say, all of those professors now have due dates throughout their online classes.

(Maggie McVay Lynch <mcvaylynch@HOME.COM>, DEOS-L - The Distance Education Online Symposium, January 25, 2001)

Question to Consider

Can you think of any e-learning related evaluation issues not covered in this chapter?

Activity

1. Using Internet search engines, locate an article that covers any of the following evaluation issues for online courses, and analyze the article from the perspectives of its usefulness in e-learning:

- Assessment of learners
- Evaluation of the instructional team
- Evaluation of administrative support
- Evaluation of learner support staff
- Evaluation of the delivery and maintenance team
- Evaluation of the management team
- Evaluation of the planning team
- Evaluation of the design team
- Evaluation of the production team
- Evaluation of the evaluation team

2. Locate an online program and review its evaluation aspects using the relevant evaluation checklist items at the end of Chapter 9.

References

Hudspeth, D. (1997). Testing learner outcomes in Web-based instruction. In B.H. Khan (Ed.), *Web-based instruction,* (pp. 353-356). Englewood Cliffs, NJ: Educational Technology Publications.

Romiszowski, A.J. & Chang, E. (2001). A practical model for conversational Web-based training: A response from the past to the needs of the future. In B.H. Khan (Ed.), *Web-based training,* (pp. 107-128). Englewood Cliffs, NJ: Educational Technology Publications.

Wheeler, S. (1999). Convergent technologies in distance learning delivery. *TechTrends, 43*(5), 19-22.

Evaluation Checklist

Evaluation of the E-Learning Content Development Process

Figure 1. People process product continuum for the content development process

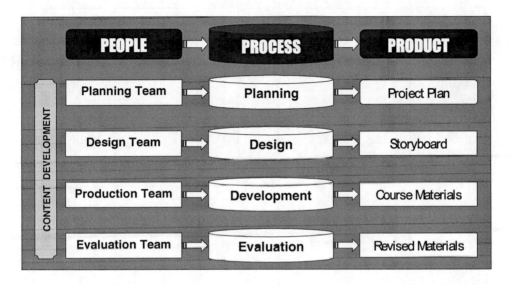

Check the overall *performance of individuals* involved in the various stages of the content development process. (check all that apply):

Role of Individual	Name of Person	Content Development Stages															
		Planning				Design				Production				Evaluation			
		Excellent	Good	Fair	Poor	Excellent	Good	Fair	Poor	Excellent	Good	Fair	Poor	Excellent	Good	Fair	Poor
Director																	
Project Manager																	
Business Developer																	
Consultant / Advisor																	
Research and Design Coordinator																	
Content or Subject Matter Expert																	
Instructional Designer																	
Interface Designer																	
Copyright Coordinator																	
Evaluation Specialist																	
Production Coordinator																	
Course Integrator																	
Programmer																	
Editor																	
Graphic Artist																	
Multimedia Developer																	
Photographer / Videographer (cameraman)																	
Learning Objects Specialist																	
Quality Assurance																	
Pilot Subjects																	

Check the *performance level of the management* team in managing the e-learning projects. (check all that apply):

Skill Types	Management Team														
	Director					Project Manager					Other				
	Excellent	Good	Fair	Poor	NA	Excellent	Good	Fair	Poor	NA	Excellent	Good	Fair	Poor	NA
Recruiting															
Supervising															
Budgeting															
Planning															
Scheduling															
Assigning tasks to team members															
Interpersonal															
Presentation															
Technological															
Research															
Outsourcing projects components															
Tracking project progress															
Conducting meetings															
Oral communication															
Written Communication															
Consensus building															
Conflict resolution															
Ability to work with others on a team															
Other (specify)															

Rate the performance level of the various *stages of the content development* process:

Content Development Stage	Performance					
	Excellent	Good	Fair	Poor	NA	Comments
Planning						
Design						
Production						
Evaluation						

Rate the performance level of the various *types of evaluation* process:

Evaluation Type	Performance					
	Excellent	Good	Fair	Poor	NA	Comments
Content Review						
Rapid Prototype						
Alpha Class						
Beta Class						
Other (specify)						

Rate the performance level of the following *tools* and *services* used during the content development process:

Tools and Services	Performance					
	Excellent	Good	Fair	Poor	NA	Comments
Content Development /Authoring Tool						
Learning Management System						
Screen Reader Software						
Accessibility Evaluation Tool						
Network Server						
Hardware Vendor Services						
Software Vendor Services						
Postal Delivery Services						
Other (specify)						

Rate the performance level of the following sites:

Site	Performance					
	Excellent	Good	Fair	Poor	NA	Comments
Development Site						
Project Support Site						
Knowledge Management Site						
Other (specify)						

Rate the *products* of the various stages of the e-learning content development process.

Stages	Product Type	Product Quality					
		Excellent	Good	Fair	Poor	NA	Comments
Planning	• *Project Plan*						
	•						
	•						
	•						
Design	• *Storyboard*						
	•						
	•						
	•						
Production	• *Course Materials*						
	•						
	•						
	•						
Evaluation	• *Revised Materials*						
	•						
	•						
	•						

Evaluation of the E-Learning Environment

Figure 2. People process product continuum for the delivery and maintenance process

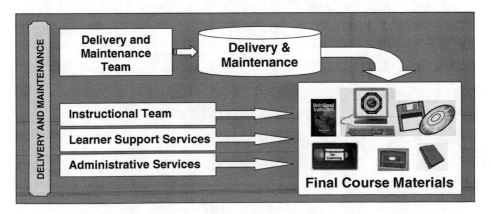

Rate the overall performance of the individuals involved in the delivery and maintenance stages of e-learning. (check all that apply):

Role of Individual	Name of Individual	Performance Level									
		Delivery Stage					Maintenance Stage				
		Excellent	Good	Fair	Poor	NA	Excellent	Good	Fair	Poor	NA
Project Manager											
Delivery Coordinator											
Systems Administrator											
Server/Database Programmer											
Other (specify)											

Rate the performance of security measures in the course:

❑   Excellent

❑   Good

❑   Fair

❑   Poor

❑   Not applicable

Rate the overall performance of individuals involved in the instructional stage. (check all that apply):

Role of Individual (Instructional Team)	Individual Name	Performance					
		Excellent	Good	Fair	Poor	NA	Comments
Online Course Coordinator							
Instructor (or Trainer)							
Instructor Assistant							
Tutor							
Discussion Facilitator/Moderator							
Learning Objects Specialist							
Copyright Coordinator							
Guest Speaker (or outside Expert)							
Other (specify)							

How are instructor's written communication skills?

❑   Excellent

❑   Good

❑   Fair

❑   Poor

How are the instructor's oral communication skills?

❑   Excellent

❑   Good

❑   Fair

❑   Poor

How is the instructor's or trainer's level of enthusiasm in teaching the course?

❑   Excellent

❑   Good

❑   Fair

❑   Poor

Rate the instructor's or trainer's performance in promoting online interaction in the course?

❑   Excellent

❑   Good

❑   Fair

❑   Poor

❑   Not applicable

Rate the overall performance of the learner support staff involved in the instructional stage. (check all that apply):

Role of Individual (Learner Support)	Individual Name	Performance					
		Excellent	Good	Fair	Poor	NA	Comments
Technical Support Specialist							
Library Services							
Counseling Services							
Customer Service							
Other (specify)							

Rate the overall performance of the administrative support services. (check all that apply):

Administrative Support	Performance					
	Excellent	Good	Fair	Poor	NA	Comments
Admission						
Registration						
Payment						
Bookstore						
Financial Aid						
Other (specify)						

Evaluation of E-Learning at the Program and Institutional Levels

Check if the institution or program has an online evaluation method (e.g., evaluation forms) in place to evaluate the performance of any of the following individuals and support services. (Note: Since several entities such as people, tools, and resources are involved in an e-learning course, the institution should develop evaluation instruments encompassing these entities. There can be a single evaluation instrument for each individual or a combined evaluation instrument for the course, program, and institution.) Check all that apply:

Individual and Support Services	Online Evaluation Method Available?			
	Yes	No	NA	Other
Director				
Project				
Business Developer				
Consultant / Advisor				
Research and Design Coordinator				
Content or Subject Matter Expert				
Instructional Designer				
Interface Designer				
Copyright Coordinator				
Evaluation Specialist				
Production Coordinator				
Course Integrator				
Programmer				
Editor				
Graphic Artist				
Multimedia Developer				
Photographer/Videographer (cameraman)				
Learning Objects Specialist				
Quality Assurance				
Pilot Subjects				
Delivery Coordinator				
Systems Administrator				
Server/Database Programmer				
Online Course Coordinator				
Instructor(or Trainer)				
Instructor Assistant				
Tutor				
Discussion Facilitator/Moderator				
Customer Service				
Technical Support Specialist				
Library Services				
Counseling Services				
Administrative Services				
Registration Services				
Marketing				
Other (specify)				

Check the status of the accreditation review by accrediting agencies:

Name of Accreditation Agency	Review Status	
	Satisfactory	Unsatisfactory

Rate the performance of e-learning marketing by the institution:
- ❑ Excellent
- ❑ Good
- ❑ Fair
- ❑ Poor
- ❑ Not applicable

Rate the effectiveness of institution's e-learning partnerships with other institutions:
- ❑ Excellent
- ❑ Good
- ❑ Fair
- ❑ Poor
- ❑ Not applicable

Does the course use an instant feedback button on most screens/pages in order to receive learners' feedback for improvement of the course?
- ❑ Yes
- ❑ No
- ❑ Not applicable

Does the course have an archive of previous students' evaluations of the course? (Note: Most online courses are modified/updated every semester based on new content and students' evaluation. Therefore, previous students' evaluation reports can sometimes be irrelevant and misleading. Professor Christopher Dede of the Learning and Teaching program at the Harvard University posts students evaluations from the previous semester.

URL: http://www.gse.harvard.edu/~dedech/502/T-5022.pdf)

❏   Yes
❏   No
❏   Not applicable

Is the course peer-reviewed during its design and development process by the following individuals? Check all that apply:
❏   Other online instructors within the institution
❏   Peers in the field
❏   Outside consultant
❏   Not applicable
❏   Other (specify)

Does the course conduct pilot testing with the target population?
❏   Yes
❏   No
❏   Not applicable

Does the course offer students an opportunity to give feedback to the institution about the quality and benefits/disadvantages of the course?
❏   Yes
❏   No
❏   Not applicable

Does the institution conduct regular surveys to find out about student satisfaction with the course?
❏   Yes
❏   No
❏   Not applicable

Do students recommend the course to others?
❏   Yes
❏   No
❏   Not applicable

Are course materials reviewed semesterly (or quaterly) to ensure their quality?

❑   Yes

❑   No

❑   Not applicable

If *yes*, check all that apply:

❑   Peer reviewed by colleagues

❑   Reviewed by the program coordinator

❑   Reviewed by the department chair

❑   Reviewed by the dean

❑   Reviewed by the course review committee

❑   Other (please describe below)

Does anyone in the institution (e.g., program director, department chairman, or training manager) review transcripts of instructor's feedback in online facilitation to evaluate his or her performance?

❑   Yes

❑   No

❑   Not applicable

❑   Other (specify)

Is the learning environment interactive? (Note: A well-designed learning environment should provide interactive experiences for it various learning tasks. When learners are engaged, they learn and enjoy.)

❑   Yes

❑   No

❑   Not applicable

Does the institution keep a record of learner completion rates every time the course is offered?

❑   Yes

❑   No

❑   Not applicable

❑   Other (specify)

Rate the quality of e-learning orientation sessions (provided by the institution) for the following stakeholder groups? (check all that apply):

Role of Individual	Quality				
	Excellent	Good	Fair	Poor	NA
Students					
Instructor					
Trainer					
Facilitator					
Tutor					
Technical Support					
Help Desk					
Librarian					
Counselor					
Facilitator					
Administrative Staff					
Other (specify)					

How effective was the technology infrastructure for the course?

❑   Very effective

❑   Moderately effective

❑   Not effective

❑   Not applicable

❑   Other

Assessment of Learners

How are the learners assessed during the course? Check all that apply:

❑   Pre-test

❑   Post-test

❑   Diagnostic test

❑   Topical/Research Paper

❑   Group Projects

❑   Individual Project

❑   Online Presentation

❑   Assignments

❑   Proctored online tests

❑   Proctored written tests

- ❏   Portfolio development
- ❏   Case studies
- ❏   Lab report
- ❏   Journal (Web logs or blogs)
- ❏   Not applicable
- ❏   Other (specify)

Are assignment types selected for the course appropriate for the content types?
- ❏   Yes
- ❏   No
- ❏   Not applicable
- ❏   Other (specify)

Check if any of the following test formats are used in the course:
- ❏   Multiple choice
- ❏   True/false
- ❏   Fill-in-the blanks
- ❏   Short Answer
- ❏   Essay questions
- ❏   Randomized quizzes
- ❏   Timed quizzes
- ❏   Quizzes with possible retries
- ❏   Scoring online
- ❏   Score analysis
- ❏   Score reporting online
- ❏   Not applicable
- ❏   Other (specify)

Rate the appropriateness of the test *format* for each learning objective:

Lesson	Objective	Number of Test Items					Length of Test Items				
		Excellent	Good	Fair	Poor	NA	Excellent	Good	Fair	Poor	NA
1	1										
	2										
	3										
	4										
	5										
	6										
	7										
	8										
2	1										
	2										
	3										
	4										
	5										
	6										
	7										
	8										
3	1										
	2										
	3										
	4										
	5										
	6										
	7										
	8										

Rate the adequacy of the *number* and *length* of test items in each learning objective:

Lesson	Objective	Number of Test Items					Length of Test Items				
		Excellent	Good	Fair	Poor	NA	Excellent	Good	Fair	Poor	NA
1	1										
	2										
	3										
	4										
	5										
	6										
	7										
	8										
2	1										
	2										
	3										
	4										
	5										
	6										
	7										
	8										
3	1										
	2										
	3										
	4										
	5										
	6										
	7										
	8										

Does the assessment provide students with the opportunity to demonstrate what they have learned in the course?

❑   Yes

❑   No

❑   Not applicable

Does the course provide students with clear grading criteria?

❑   Yes

❑   No

❑   Not applicable

❑   Other (specify)

What is the timeline for the instructor to provide feedback on assignments? (Note: The feedback timeline will depend on the type of assignment.)

❑ Immediate

❑ Within a week of the assignment's receipt

❑ Within two weeks of the assignment's receipt

❑ Within a month of the assignment's receipt

❑ Open time

❑ Not applicable

Does the course give sufficient time for student to complete course assignments?

❑ Yes

❑ No

❑ Not applicable (e.g., self-paced)

Does the course require students to log on to the course Website during certain periods of time (e.g., at least once a week) as proof of their attendance?

❑ Yes

❑ No

❑ Not applicable

If *yes*, does the instructor or facilitator contact the absent students?

❑ Yes

❑ No

❑ Not applicable

Does the course require students to participate in online discussions?

❑ Yes

❑ No

❑ Not applicable

If *yes*, how (check all that apply):

❑ By responding to main discussion topics/questions posted by the instructor or facilitator

❑   By responding to other students' posting for the original topic/
question
❑   Other

Does the course clearly explain how a student's discussion participation contributes toward a student's grade?

❑   Yes
❑   No
❑   Not applicable

Are the quizzes/tests in this course accurate and fair?

❑   Yes
❑   No
❑   Not applicable

Are the course test items re examined and evaluated to identify questions
answered by all learners or questions not answered by any learners? (Note: Item
analysis is a technique to refine test questions. It is not worthwhile to test learners
on content they already know or on content that none of them can answer. Test
items that are poorly designed can be confusing to learners and therefore
unanswerable.)

❑   Yes
❑   No
❑   Not applicable
❑   Other (specify)

Are the test questions (or items) refined after analyzing learners' performance?

❑   Yes
❑   No
❑   Not applicable
❑   Other (specify)

Who grades students' assignments? Check all that apply:

Assignment Types	Graded By				
	Instructor	Tutor	Computer	Other	Not Applicable
Quiz					
Test					
Project					
Discussion forum postings					
Journal reports					
Paper					
Other					

Does the course accept late assignments?

❑   Yes

❑   No

❑   Not applicable

If *yes*, is there any penalty in the form of reduced points for late assignments?

❑   Yes

❑   No

❑   Not applicable

Are students required to submit any one of the following for late assignments?

❑   Reason for late submission (system difficulties, resources not available, etc.)

❑   Doctors statement (if health related)

❑   None

❑   Not applicable

Does the instructor/tutor help students work out a plan to complete their late assignments?

❑   Yes

❑   No

❑   Not applicable

❑   Other (specify)

Are due dates adjusted for students in different time zones?

❑   Yes

❑   No

❑   Not applicable

❑   Other (specify)

Does the course provide an online testing facility?

❑   Yes

❑   No

❑   Not applicable

> If *yes*, does the online testing facility include multimedia attributes (e.g., test items are capable of including audio, video, image, etc)?
>
> ❑   Yes
>
> ❑   No
>
> ❑   Not applicable
>
> ❑   Other (specify)

Check all that apply for the course related exams and tests:

Exam Type	Exam and Test are Given					
	Weekly	*Monthly*	*Mid-Term*	*Term Final*	*Based Learners' Readiness*	*Other*
Take home						
Proctored exam						
Proctored online exam						
Online exam						
Online quizzes (Unannounced)						
Other						

If exams are proctored, how are proctors selected?

If exams are proctored, how are students identification established?

Evaluation types used in the course
- ❑   Self evaluation
- ❑   Peer evaluation
- ❑   Group evaluation
- ❑   Instructor
- ❑   Other
- ❑   Not applicable

Does the institution have policies and guidelines regarding the assessment of students that the course instructor must follow?
- ❑   Yes
- ❑   No
- ❑   Not applicable
- ❑   Other (specify)

Does the course set clear assessment standards?
- ❑   Yes
- ❑   No
- ❑   Not applicable

Does the course use a pretest to assess learners' pre-requisite skills on learning tasks in its various lessons?
- ❑   Yes
- ❑   No
- ❑   Not applicable
- ❑   Other (specify)

Does the course provide practice items for learners on its various lessons?

☐   Yes

☐   No

☐   Not applicable

☐   Other (specify)

Does the course provide frequent confirmational and corrective feedback?

☐   Yes

☐   No

☐   Not applicable

Does the course provide remedial activities?

☐   Yes

☐   No

☐   Not applicable

Does the course have a mechanism in which a learner can be truly measured and not cheat?

☐   Yes

☐   No

☐   Not applicable

Does the course provide an environment that rewards individuals or individual work?

☐   Yes

☐   No

☐   Not applicable

Does the course include authentic assessment strategies to evaluate real-world skills?

☐   Yes

☐   No

☐   Not applicable

Are assignments relevant to course objectives/goals?

❑   Yes
❑   No
❑   Not applicable

Are exercises relevant to course objectives/goals?

❑   Yes
❑   No
❑   Not applicable

Does the course provide clear instructions for preparing and submitting assignments?

❑   Yes
❑   No
❑   Not applicable

Does the course penalize participants who do not turn in their assignments on time?

❑   Yes
❑   No
❑   Not applicable

Does the course provide the option for learners to receive graded papers or assignments electronically?

❑   Yes
❑   No
❑   Not applicable

    If *no*, check all that apply:

    ❑   The instructor sends graded assignments to learners via regular mail
    ❑   The instructor sends graded assignments to learners if learners have already sent postage-paid envelopes to the instructor
    ❑   The instructor faxes graded assignments to learners
    ❑   The instructor faxes graded assignments to learners if learners pay the cost

    ❑   Learners can pick up graded assignments from the instructor's location

    ❑   Other (specify)

Does the course have a system for keeping records of student progress online?

❑   Yes

❑   No

❑   Not applicable

Does the course have a system for providing student grades online?

❑   Yes

❑   No

❑   Not applicable

Does the course clearly indicate assignment due dates?

❑   Yes

❑   No

❑   Not applicable

    If *yes*, are they suited for geographically diverse time-zones?

    ❑   Yes

    ❑   No

    ❑   Not applicable

Can students appeal any marks or points given for quizzes, essays, exams or assignments that affect their final grade?

❑   Yes

❑   No

❑   Not applicable

Activities

1. Suppose you are working for an institution that is planning an e-learning initiative. You have been asked by your institution to develop a position paper with an overview of the comprehensive e-learning process. This paper should help your institution to see the e-learning process from a birds-eye view and provide the realities of an e-learning environment. Let's call the position paper an *e-learning plan*. In developing the e-learning plan, you should consider including as many critical issues as possible encompassing the eight dimensions of the E-learning Framework discussed in the book. In the book, issues within each dimension of the E-learning Framework are presented as *questions* that course designers can ask themselves when planning, designing, developing, implementing and evaluating e-learning and blended learning materials.

    Based on your understanding of the e-learning process and items included in this chapter, develop an **e-learning plan** for your institution. The following is a sample outline for an e-learning plan:

Sample Outline for an E-Learning Plan

I. E-Learning Environment

    In this section, you will rationalize that e-learning is a viable method of providing education and training to learners dispersed all over the world. Therefore, for your target audience, you should:

    *   <u>describe</u> e-learning in your own words
    *   <u>identify</u> similarities and differences between e-learning and the traditional classroom, and
    *   <u>list</u> the advantages of e-learning over the traditional classroom instruction.

II. Institutional Issues

    In this section, you begin with rationalizing the need for an e-learning initiative at your institution and its potential benefits. Then, discuss the following institutional issues whenever applicable (e.g., if your design plan is for a corporate setting, then "financial aid" may not be an issue, whereas for an academic setting it is an important support service issue).

- ❑ Administrative Affairs
  - ❑ Budgeting and return on investment
  - ❑ Information technology services
  - ❑ Instructional development and media services
  - ❑ Marketing, admissions, graduation, certification and alumni affairs
  - ❑ Organization and change (diffusion, adoption and implementation of innovation)
- ❑ Academic Affairs
  - ❑ Faculty and staff support
  - ❑ Instructional affairs
  - ❑ Workload, compensation and intellectual property rights
- ❑ Student Services
  - ❑ Pre-enrollment services
  - ❑ Course and program information
  - ❑ Orientation
  - ❑ Advising
  - ❑ Counseling
  - ❑ Financial Aid
  - ❑ Registration and payment
  - ❑ Bookstore
  - ❑ Library support
  - ❑ Social support network
  - ❑ Tutorial services
  - ❑ Internship and employment services

III. Technological Issues

Technology issues should include:

- • Infrastructure planning
- • Assessment of institution's existing technologies and technology plan
- • Standards, policies, and guidelines related to hardware, software and other relevant technologies required for e-learning

- Software Requirements

  The tables below are provided to help you in completing your E-learning Plan. You may want to use these tables, or altered versions of them, as you work on your project and even as part of your final report.

	Software Name	Required (Req) Or Recommended (Rec)?					(List specific tasks performed by software)	Cost
		Learner	Instructor	Tech Support	Institution	Other		
Word processor								
Email package								
Presentation program								
Spreadsheets								
Database								
Authoring tools or LMS*								
Discussion software								
Operating system								
Plug-ins								
Browsers								
ASP								
AV Streaming								
Other								

*   *Learning Management System. Indicate whether LMS is SCORM or IEEE compliance.*

- Hardware Requirements

  The following tables are provided to help you in completing your E-learning Plan. You may want to use these tables, or altered versions of them, as you work on your project and even as part of your final report.

CPU										
RAM										
ROM										
Hard disk			gigabyte							
Disk drive										
CD-ROM				24x, 32x						
SDRAM	32/64/128 /256MB									
Sound card										
Speaker										
Microphone										
Video card										
DVD										
Ethernet										
Dial-in modem				28.8, 33.6. 56 Kbps						
DSL										
Cable modem										
Wireless Internet connection										
Monitor						12" 14" 16" Other	640X480 800X600 1024X768 256, thousands, millions			
Ink-jet										
Laser										
Digital camera										
Video camera										
Other										
Comments										

Note: Type L=Learner, I=Instructor, T=Tutor, S=Technical Support and IN=Institution wherever applies in the above Table.

IV. Pedagogical and Evaluation Issues

- Discuss the instructional approach (unstructured vs. structured learning activities) for designing the course content. Here, you can describe how the overall design of learning activities for various parts of the course content can be developed: highly structured, mostly structured, loosely structured or unstructured.

- Provide a brief description of at least five of the following instructional methods emphasizing how they are used successfully in courses (provide course URLs).

  Presentation

  Demonstration

  Drill and Practice

  Tutorials

  Games

  Story Telling

  Simulations

  Role-playing

  Discussion

  Interaction

  Modeling

  Facilitation

  Collaboration

  Debate

  Field Trips

  Apprenticeship

  Case Studies

  Generative Development

  Motivation

- Discuss how learner assessment will be designed?

- Discuss how instructor evaluation will be conducted?

- Discuss how the design of the learning environment will be assessed?

V.   Interface Design and Ethical Issues

- Discuss interface design issues including site design, navigation and usability testing for e-learning.

- Discuss ethical considerations that should be taken into account in designing e-learning; including, social and cultural diversity, bias, geographical diversity, learner diversity, information accessibility, etiquette and legal issues (e.g., policies and guidelines, privacy, plagiarism and copyright).

VI. Resource Support and Management issues

- Discuss online support services including instructional, counseling and career guidance.

- Discuss both online and offline resources available for e-learning.

- Discuss the maintenance of e-learning sites and the distribution of information.

VII. E-Learning Case Studies

- Using the Internet's search engines, identify and write about two or three existing e-learning programs that encompasses many of the issues covered in the above outlines.

2.   Many organizations including corporations, government agencies, nonprofits, and educational institutions are currently using e-learning and blended learning materials for their various educational and training programs. Using the checklist items in this chapter, conduct a program evaluation of an online program.

Reference

Boshier, R., Mohapi, M., Moulton, G., Qayyum, A., Sadownik, L. & Wilson, M. (1997). Best and worst dressed web courses: Strutting into the 21st century in comfort and style, *Distance Education, 18*(2), 327-348.

About the Author

Dr. Badrul H. Khan is associate professor of education and director of the Educational Technology Leadership graduate cohort program at The George Washington University. He is founder of *BooksToRead.com*, a recommended readings site on the Internet. Previously, he was an assistant professor of education and founding director of the educational technology graduate program at the University of Texas at Brownsville. He also served as an instructional developer and evaluation specialist in the School of Medicine at Indiana University, Indianapolis. He earned a BA in chemistry and a PhD in instructional systems technology from Indiana University, Bloomington, IN.

While growing up in Chittagong, Bangladesh during the 1970s, Dr. Khan used to dream about having access to the well-designed learning resources available only to students in industrial countries. In the '70s it was unthinkable to have equal access to those resources. In the '90s, with the emergence of the World Wide Web, Khan's dream of equal access to quality learning resources became a reality. His desire for broadly available distributed learning systems and his scholarly grounding in the field of educational systems design and technology have enabled him to present a total vision for educational and training possibilities of the new worldwide communications technologies.

Through his teaching and publishing, Dr. Khan has been instrumental in creating a coherent framework for Web-based instruction, training, and learning. In his first book, *Web-Based Instruction* (Educational Technology Publications, 1997), he took a leadership role in defining the critical dimensions of this new field of inquiry and practice at all levels of education. Reflecting its enormous acceptance worldwide, *Web-Based Instruction* has become a bestseller and has been adopted by colleges and universities worldwide. His second book *Web-Based Training* (Educational Technology Publications, 2001) is a landmark book that covers all aspects of Internet's World Wide Web for training at all levels.

He continues to advance the discourse in the field of distance learning. His contribution to the field of open, flexible and distributed learning is recognized throughout the world. As a result, the following books based on his e-learning framework are published: *Managing E-Learning* (Idea Group Publishing, USA), *E-Learning Strategies* (Seohyunsa Press, Korea; and Erickson, Italy*), Implementing E-Learning* (Dar Shua' Printing and Publishing, Syria*), E-Learning: Design, Delivery and Evaluation* (Beijing Normal University Press, China*), and E-Learning Quick Checklist* (Idea Group Publishing, USA). Dr. Khan's e-learning books are translated into several languages.

He is currently working on a new book, *Flexible Learning* (in press, Educational Technology Publications, USA), which will include case studies, design models, strategies, and critical issues encompassing the multiple dimensions of his e-Learning Framework. His framework is being recognized as a model for distance learning by the publications of Commonwealth of Learning (COL), an organization of commonwealth countries.

A sought-after keynote speaker on Web-based instruction and elearning, Khan is past president of the international division of the Association for Educational and Communication Technology (AECT). He delivered keynote addresses at the various distance learning conferences organized by the ministry of education in Turkey, China, UAE, Bahrain, Oman and Saudi Arabia; and academic and professional conferences in the USA, Canada, Korea, India and Bangladesh. He was one of the select few experts invited to a symposium on virtual education organized by the White House Office of Science and Technology Policy (OSTP). He advised the EdTech Team at the US Department of Defense for the *Joint Professional Military Education in 2010: The EdTech Report*. He provided advisement in elearning related issues for the development the *National Education Technology Plan* for the U.S. Department of Education. He served as a consultant/advisor to distance education related projects at the World Bank and Ministry of Education in several countries, and academic institutions and corporations in the USA and abroad. Dr. Khan interviews visionary leaders in technology-based education for a regular section of the *Educational Technology* magazine entitled *Interviews with Badrul Khan*.

He is a contributing editor of *Educational Technology* (USA*)*, a consulting editor of *The International Review of Research in Open and Distance Learning* (Canada*)*, a member of the editorial board of *Distance Education Journal* (Australia) and a member of the editorial board of *International Journal of Learning Technology* (UK), a member of the advisory board of

Media and Technology for Human Resource Development, a member of the advisory board of *Indian Journal of Training & Development,* a member of editorial advisory board of *the eLearning Digest* (UAE), a member of the editorial board of *eLearning* (Italy) and a member of the advisory board of *Review of Education at Distance* (Brazil). Dr. Khan's homepage is available at: BooksToRead.com/khan.

Index